Science and Technology
in a Multicultural World

Science and Technology in a Multicultural World

THE CULTURAL POLITICS OF FACTS AND ARTIFACTS

David J. Hess

Columbia University Press

NEW YORK

Columbia University Press
New York   Chichester, West Sussex

Library of Congress Cataloging-in-Publication Data
Hess, David J.
    Science and technology in a multicultural world: the cultural politics of facts and
artifacts / David J. Hess.
    p.   cm.
    Includes bibliographical references and index.
    ISBN 0-231-10196-1
    ISBN 0-231-10197-X (pbk.)
    1. Science—Social aspects.   2. Technology—Social aspects.
3. Multiculturalism.   I. Title.
Q175.5.H47   1994
303.48'3—dc20   94-34733
                                                                     CIP

Printed in the United States of America

c 10 9 8 7 6 5 4 3 2 1

Contents

Preface

Like it or not we are living in a *multicultural world*. I use that term in a deliberately ambiguous way to signal various aspects of the contemporary (or postmodern) condition. To begin, it is a multicultural *world*. Developments in transportation and communication as well as the crisis of world ecology have created the so-called global society. Furthermore, as the globe has shrunk in size it has also become a *multicultural* world. People of diverse nationalities find themselves in increasing contact with each other. In many countries women, underrepresented ethnic groups, gays and lesbians, and other previously excluded groups have gained a greater voice in government, the media, and the professions. Finally, the cultural world in which people are living is marked by public debates on diversity, pluralism, oppression, exclusion, inclusion, colonialisms, identity politics, and other issues that can be glossed as multicultural.

For many people involved in the debates the very word *multiculturalism* is controversial. From the right critics worry that increasing attention to identity politics will undermine citizenship, patriotism, and other mechanisms of national stability and integration. From the left critics worry that the attention to culture occludes racism or patriarchy by reducing them to a form of ethnocentrism, or that the official appropriations of multiculturalism represent a new form of political pluralism that glosses over conflict and domination. From either perspective the term *multiculturalism* is itself problematic if not altogether wrongheaded.

I am more sympathetic to the criticisms from the left than those from the right. Those on the right, I believe, worry about something that is not happening, for they miss the complexities of how national cultures continue to reproduce themselves in new settings—including, for example, national styles of multiculturalism or identity politics. However, unlike some critics on the left, I am not willing to censor either the word *multiculturalism* or the value of a cultural perspective. I see multiculturalism less as a monolithic social phenomenon than as a social and ideological space that has opened up new possibilities for maneuvering and social change. Because I see multiculturalism as a cultural space, a stage on which conflicts are taking place and history is being made, I disagree with appropriations of the word that would make it into the benign, conflict-avoiding version of liberal pluralism that one finds among many corporate diversity trainers. I do not assume that the dramas on this stage among different classes, genders, nationalities, races, ethnicities, and so on are always or even mostly benign conflicts. This book is not going to be a story of smiling, Disneyfied faces singing a happy multicultural song about how small and diverse the world is.

In the United States the idea of multiculturalism is often associated with debates about curriculum diversity, and as an educator that issue is also very close to my heart. For the most part curriculum discussions have focused on the literary canon, which in Western societies has been dominated by men of European descent, often from the privileged classes. Whether one is a canon defender, canon debunker, or an advocate of some combination, it is no longer possible to compose a syllabus in the humanities without considering questions of balance, bias, and the inclusion of historically excluded voices. However, such is generally not the case for introductory surveys of engineering, medicine, and the natural sciences—or for surveys of the history, philosophy, and social studies of the technical fields. Those courses generally still assume a canon that runs from Aristotle, Galen, and Ptolemy through Copernicus, Galileo, Harvey, and Newton, and on to Darwin, Pasteur, and Einstein. The story of what constitutes international science and technology today is largely limited to the viewpoint of the experts who are seen to have produced their fundamental principles, and historically those experts have been generally middle-to-upper-class men who were European or of European descent.

In this book I take a step toward puzzling through a different approach to science and technology, one that takes as its starting point the culture concept and the question of how to include historically excluded perspectives. The great strength of the cultural approach to issues of science and technology is in challenging people to consider how other cultures and groups may have new and different ways of defining the true and the useful. My perspective is informed by an anthropological concept of culture; that is, I view culture as a symbolic apparatus that underlies a wide range of ideas and practices. However, like many other anthropologists today I believe that the culture concept must be flexible, sophisticated, and above all not naive. It is fine to talk about diversity and differences, but usually the people on one side of the difference have the upper hand and want to impose their way on the other side. Thus, cultural analysis must go hand in hand with attention to questions of power.

The topic of multiculturalism, science, and technology is timely because today's students and faculty, administrators and alumni, educators and noneducators all recognize the importance of science and technology in the global economy. They also recognize that scientific and technical occupations in the United States and other countries are becoming increasingly diverse and internationalized. Many students—including many white males—now studying to enter the technical occupations are hungry for course materials on gender, race, ethnicity, non-Western cultures, and diversity issues in general as they relate to science and technology. My experience of this general interest is firsthand, because for several years I have been teaching courses on the topic. In the process I have been puzzling through the problem of teaching about power and domination issues without falling into the trap of science-and-technology bashing.

One of the frustrations I have encountered in my teaching is that much of the information on the topic is scattered through a wide number of journals and books, many of which are difficult to locate or understand. I therefore have written this book to make available a resource for a wide variety of research on technoscience, culture, and power. I have written it to be accessible to nonspecialists such as advanced undergraduates, graduate students, science teachers, scientists, engineers, or other members of the reading public. In the role of a guide, I have also gone beyond summaries and provided my own syntheses of research findings.

There are already several introductory texts to the field of STS, or science and technology studies (e.g., Webster 1991, Woolgar 1988a) as well as histories of non-Western science and technology (e.g., Ronan 1982,

Pacey 1990) and collections of essays on science, race, and gender (e.g., Harding 1993, Tuana 1989). My goal is to move questions of culture and power closer to center stage of the rapidly developing interdisciplinary dialogue of STS by synthesizing some of the research. Because my home discipline of cultural/social anthropology has addressed many of the questions with which I am concerned, many of my examples come from that field. My aim, however, is not to slight other disciplines; rather, it is to show some of the contributions from a perspective rooted in my home discipline. I also draw on research in the history of science, intellectual history, intercultural communication, the sociology of science, cultural studies, and development studies. I would like to think of this book as a kind of impressionist landscape painting; specialists in each area may find some gaps in my discussions, but I hope that what may be lost in depth is made up for in breadth. I write this book in the belief that there is a role for generalists, for those who wish to see and save the forests of the social sciences and humanities, for those who wish to resist the disciplining of the disciplines.

Acknowledgments

I would like to thank the Instructional and Development Program at Rensselaer Polytechnic Institute, which provided me with a faculty development grant that allowed me to begin research on this project. My thinking also developed in the many exchanges with students and faculty in Rensselaer's Science and Technology Studies Department. I also benefited from discussions in the university's Committee on Diversity and Multiculturalism as well as the FIPSE-sponsored faculty research group dedicated to cross-cultural studies of science and technology in Rensselaer's School of Humanities and Social Sciences. Likewise, the annual conferences on Comparative Science and Technology, organized by members of the Massachusetts Five-College consortium, have been very helpful. I had the privilege of organizing the third Comparative Science and Technology conference at Rensselaer, and I benefited from discussions at that conference as well.

My project is only one book in a rapidly growing interdisciplinary field of cultural and anthropological studies of science and technology. I have developed my ideas in dialogue with a community of anthropologists and feminist/cultural studies researchers, many of whom are mentioned in the pages that follow. We have exchanged ideas in lively sessions on science, technology, culture, and power at the Society for Social Studies of Science, on cyborg anthropology at the American Anthropological

Association, and at the School for American Research seminar, also on cyborg anthropology. Colleagues at those meetings provided useful feedback on earlier versions of some of the chapters, as did my hosts for invited colloquia at the Wesleyan University Center for the Humanities and Cornell University's Science and Technology Studies Department. I wish to thank specifically Kathy Addelson, Diana Forsythe, Shirley Gorenstein, Deborah Johnson, John Koller, Linda Layne, Brian Martin, Roxanne Mountford, Paul Rabinow, Sal Restivo, John Schumacher, Ray Stokes, Sharon Traweek, and Jim Zappen for reading parts or all of the manuscript and providing me with comments and additional bibliographic references.

Science and Technology
in a Multicultural World

1 | Introduction

Almost every day a news report tells of another innovation in science and technology. In some cases the report also discusses ethical dilemmas, political conflicts, or economic impact associated with the innovations. At a popular level, then, there is already quite a bit of recognition of the interwoven contours of science/technology and society/culture. Nevertheless, thinking about the issue is often muddled. It can therefore be helpful to look at what researchers of science and technology—principally anthropologists, historians, philosophers, political scientists, and sociologists—have to offer.

Let me begin with some definitions. I will understand science to be knowledge about the natural world, just as social science is knowledge about the social world and technology is materially embedded knowledge about how to create effects via artifacts (that is, via "things" such as machines, tools, and architecture). In modern societies science is institutionalized, and therefore the term *science* includes institutions, networks, and other social aspects associated with the production of scientific knowledge. Knowledge production involves finding patterns in observations, and it is organized into explanations of those observations. Historically (and in some cultures today) those explanations are sometimes of an otherworldly or mythological character, but in the modern world they take the form of theories that usually rely on assumptions of materialism and mechanism.

The interdisciplinary study of science and technology is currently known as STS, or science and technology studies. (Sometimes the third *S* in STS is for "society": science, technology, and society studies.) Since the 1970s the interdisciplinary field has grown enormously. Universities across the world have realized the importance of bridging the technical and humanistic professions by setting up STS programs and departments. In the process STS has taken on a life of its own, and it has introduced some at times counterintuitive new concepts. The fundamental starting point for STS today is the idea that theories, observations, methods, machines, social relations, institutions, networks, and other aspects of the technoscientific world are in some sense socially shaped, negotiated, or otherwise "constructed." The term *social construction* has a variety of uses and meanings, so it makes some sense to begin by defining it and other related terms, particularly the related concept of cultural relativism.

Social Constructivism and Cultural Relativism

Popular accounts sometimes assume that science and technology are produced according to purely rational (or cognitive) factors. The actual making of science and technology is assumed to be in some sense outside society and culture. If social factors enter the production process, it is in the form of corruption, bias, or influence by interests of some sort. In contrast to this popular model mounting evidence in STS has shown that social factors shape the production of all science and technology. In other words, the production or making of science and technology—their construction—is a social process. Even the very cognitive assumptions behind what constitutes a rational argument, evidence, and so on are rooted in the beliefs of a specific discipline or scientific community. Thus, even the cognitive or rational can be seen as sociocultural.

To explain the concept of social constructivism through example, it can be said to include the following:

1. Decisions. What knowledge is produced and how it is to be used are socially driven decisions. The proposition is fairly obvious

considering the political nature of the allocation of research funds or the strategic nature of scientists' decisions about what research project is worth undertaking. Social factors clearly play a large role in the direction of the growth of knowledge, the drawing of boundaries between acceptable and unacceptable research, and so on.

2. Observations. What people expect to observe, are able to observe, and want to observe are all shaped in part by their theories and assumptions, which in turn are outcomes of discussions and controversies in which social negotiation is critical. Even apparently transparent observations, such as machine inscriptions of data, are social because machine design is the product of a history that involves social negotiation, as are decisions over calibration and how to interpret machine inscriptions. (However, as will be explained below, this claim does not mean that observations have nothing to do with reality: observations are simultaneously socially shaped and representative of a "real" material or social world.)

3. Generalizations and theories. A number of STS studies has documented as social processes such key scientific and technical events as the eruption of controversies, consensus formation, rates of acceptance, assessments of credibility, the conversion of knowledge claims into widely accepted facts, the acceptance and rejection of theories, and so on. It is possible to distinguish between true and false knowledge, but the criteria for making those distinctions cannot be derived from an abstract and universal faculty of reason. Rather, they are historically situated in specific disciplines and scientific communities.

4. Methodology. Decisions on appropriate methods, criteria for establishing replication, statistical measures, quantitative versus qualitative measurement, and so on are shaped by rhetoric, network politics, disciplinary cultures, personal reputations, gender socialization patterns, and so on. There is no single Scientific Method to which all scientists can refer; instead, laboratory procedures are opportunistic and contingent on social factors.

The culture concept often enters into discussions of social construction in the form of the corollary that different cultures have different sciences and technologies. Today's modern, international science and technology—the kind that is taught in most universities throughout the world—is a product of today's modern culture in the same way that ancient Greek science, Hindu science, Chinese science, and so on were products of their cultures. Modern international science is often called *Western* science, a term that belies its cultural roots. A second corollary is the idea that today's international science sometimes varies across cultures, genders, or ethnic groups. In other words, in at least some cases it is possible to locate different "styles" of science and technology.

I will argue that the culture concept has a much more important role to play in the social studies of science and technology than as a corollary to the principle of the social construction of knowledge. This book will explore a number of other contributions of the culture concept to STS research, and in the process I will argue for an extension of social constructivism that is based on a cultural perspective as understood by anthropologists: a cultural constructivism. First, however, it is important to clarify some terms in order to avoid misunderstandings.

Misunderstandings Regarding the Term Relativism

A cultural perspective begins with the interpretation of the meaning of science and technology from the perspective of different communities: expert producers, yes, but also nonexperts. As I shall demonstrate in this book, a cultural perspective results in some fairly profound changes or extensions of STS research. The art and principle of interpreting the meaning of science and technology *relative* to the understandings of local communities—either expert producers or nonexpert users—is sometimes known as cultural relativism. I will restrict the scope of the term *cultural relativism* to this methodological principle, which is widely accepted by cultural anthropologists as well as many historians and sociologists. Although the principle reflects a starting point in the methods of many researchers, their explanatory and interpretive accounts are not limited to local understandings.

Unfortunately, some people use the term *cultural relativism* as a synonym for other kinds of relativism. It is important to clarify the confusions so that readers do not misinterpret either social constructivism or cultural relativism. I divide the other relativisms into three main types: epistemological, metaphysical, and moral.

The first criticism is that either social constructivism or cultural relativism involves some form of what I will call epistemological relativism. For cultural relativism the argument starts by stating that some social scientists begin by interpreting knowledge with respect to what it means to local communities, and therefore they assume that each community has its own version of the truth. In this sense knowledge is "relative" to the community. Therefore, the argument proceeds, all knowledge is relative and there is no way of deciding between true and false knowledge. A tribe in the Amazon may never hear of the laws of modern physics, or a New Age commune in California may not believe in them. Their views about physics are, according to cultural relativists, equally valid as the views of physicists.

The argument is misleading because anthropologists and other researchers who use cultural relativism as a principle of research methodology do not stop with local interpretations. It is true that we try to suspend temporarily our own assumptions about the world in order to understand how other people see it. However, we still make judgments about whether or not other interpretations are valid. Except for social scientists who go native and are converted to other world-views, we generally make those judgments based on the knowledge and methods of the intellectual communities in which we circulate. Thus, it *is* possible to distinguish true from false knowledge, and we do it all the time.

For social constructivism the parallel argument is that because theories are socially constructed, only social factors determine the truthfulness of theories, and so it seems there would be no way of distinguishing between true and false theories. Therefore, one criterion for truth is as good as another. The reply to the argument is that theories can be shaped by social factors but simultaneously they can be accepted as "true" by a given community of scientists. Thus, it is not the case that only false knowledge is socially constructed. Knowledge deemed to be true is as equally social as knowledge deemed to be false (an idea known as the symmetry principle). Furthermore, social constructivists can—and usually do—believe in some criteria for deciding which knowledge claims are more or less credible. However, they do not think that the criteria are derived from a transcultural, universal faculty of reason. Rather, most supporters of constructivism and/or cultural relativism hold that what counts as reasonable varies from one culture, discipline, and time period to another. Certainly, in order to be able to make any arguments at all, I will have to assume standards of evidence and argumentation that are shared by the disciplinary cultures of today's social sciences and humanities. Yet, while I assume that those standards of evidence and argumentation produce relatively accurate maps of observations of the social world, I also recognize them to be changing products of a long history in which ideas about rationality are embedded in social processes and cultural assumptions. As a member of several social science communities at a given point in their histories, I have my criteria for deciding truthfulness, and I believe they are right. However, I would also reject anyone's claim that there is an asocial or transcultural way of justifying our beliefs.

A second argument against cultural relativism and social constructivism is the confusion with what I will call metaphysical relativism. An example of this kind of argument is the following: 1. some social scientists believe that observations are shaped by social factors or structured by cultural categories; 2. they think that observations are *only* shaped by the social or cultural; and 3. therefore, they think that the world beyond observations—the out-there of reality—does not exist. Instead, the only reality is the social/cultural, which is projected onto observations. Constructivists or cultural relativists, therefore, believe in a form of social idealism. They do not believe in a real, material world beyond the data of observations.

The argument may be true for some social scientists, but it is not necessarily true. Observations are shaped by social factors or structured by cultural categories, but they are, at the same time, shaped and structured by an external reality. Most social scientists are like natural scientists in that both assume a real, material world beyond their observations. A metaphor might help. (Of course, whenever someone says a metaphor is coming, my cultural antennae go out. Aware as I am in my ongoing dialogue with my partner that some men navigate by maps and some women by landmarks, I am not at all unconscious of the masculinity of the metaphor that follows. Also, as an anthropologist who in Brazil learned not to open a map in public, I am aware that using maps is a very culture-bound phenomenon. At a metalevel, however, the cultural meanings of the metaphor only proves the larger point of the inescapableness of the social-cultural.) Here is the metaphor: natural and human scientists alike make maps of territories (the real) that we never see directly, only through our lenses. What we see is inflected by previous maps, our mapmaking skills, the types of lenses we wear, and what we want to find, all of which are imbued with the social. Not everyone ends up looking at the same territory or even drawing the same map. What is left out of the map becomes as significant as what goes in it.

Furthermore, when natural scientists today observe reality, their observations are more complicated than looking at leaves or listening to bird calls (although even categories of plants and animals vary significantly across cultures). Rather, the physical world is usually mediated through machines and their inscriptions, which are subject to interpretation that in turn relies on the shared understandings of a scientific culture. In summary, as a social scientist I assume that I am producing or constructing knowledge about a real social world (one that includes scientific knowledge and technology). Certainly others' observations of that social world constrain what I am able to say credibly in the same way that intersubjective observations of natural reality constrain what scientists are able to say about the natural world. Social scientists and natural scientists alike are in the business of making observations about their realities and constructing maps of what they have seen. In short, social constructivism and cultural relativism can be perfectly consistent with realism, but reality is always apprehended through lenses.

A third type of argument is that social constructivism and cultural relativism imply moral or ethical relativism. However, this line of argument confuses statements about what is the case with statements about what ought to be the case. Philosophers refer to attempts to derive "ought" from "is" as a logical fallacy known as the naturalistic fallacy. As a methodological principle cultural relativism does not necessarily imply any moral position, from moral absolutism at one extreme to moral relativism at another. Some cultural relativists may be moral or ethical relativists, but others may adhere to a belief in fundamental human rights.

Finally, there is also a pragmatic argument in favor of social constructivism and cultural relativism. From a cultural-political perspective the framework has been enormously useful as a way of undermining naive beliefs in expert scientific authority and blind faith in technology as a way to solve social problems. In this sense social constructivism can be linked to attempts to democratize the production of science and technology and, through them, society. However, democratization goes beyond questioning philosophies of naive realism and the authority of experts. Unless democratization includes a multicultural perspective, social constructivism is very limited. This is why I join others who advocate an additional step: a culture-and-power perspective.

From the New Sociology of Science to a Culture-and-Power Perspective

The principle of the social construction of knowledge and technology was not widely accepted until the 1970s. Before that point social scientists tended to focus on institutions and leave untouched the "content" of science and technology (that is, theories, methods, facts, or technical design). The institutional focus was so closely associated with the sociologist Robert Merton that the adjective *Mertonian* is sometimes used to describe this perspective. Among historians "internalist" scholars did focus on the ideas of science, but their studies generally traced ideas to other ideas and wrote social factors out of the history. During the 1970s and 1980s a number of different programs, schools, and frameworks emerged, all of which took as a starting point some version of the idea that STS involved the study of the social content of science and technology.

I entered STS in the 1980s and found a lively set of theoretical debates going on. To a large extent much of the excitement that I found expressed in the meetings and journals of the social science wing of STS centered on what is sometimes called the "new sociology of science" or "social studies of knowledge." As I attended the meetings, read the journals, and, above all, talked with the STS researchers, it became clear to me that there was a fairly coherent group of social scientists—mostly European, mostly men (with some exceptions, such as Karin Knorr-Cetina), and mostly young enough not to remember World War II—who talked to each other, debated each other, cited each other, and, in short, formed an inner circle. They viewed themselves as epistemological radicals, for the Young Turks saw themselves as undermining the very core of the ideology of modern science: the assumption that theory, methods, and scientific knowledge in general somehow transcend the social.

Although I have learned a great deal from the new sociologists of science (and later of technology), I became increasingly unhappy with the boundaries they placed around the idea of construction. For example, the research topic of injustices or cultural differences in the social structure of scientific communities does not have much to do with the new sociologists' principal area of interest: the construction of scientific knowledge or technical design. As a result the study of social structure takes on a passé flavor. The new sociologists might even label researchers in this area "Mertonian," and therefore contribute to the construction of this kind of research project as having a backwater status.

I found other researchers who shared and confirmed my sense that many of the new sociologists of science tended to exclude as theoretically uninteresting issues involving social difference and social justice. Feminists, Third World intellectuals, scholars of historically oppressed ethnic groups and races, lesbian and gay scholars, cultural studies researchers, cultural anthropologists, radical STS scholars, and others did tend to talk about those issues. Our concerns, which I flag under the rubrics of "culture and power" or "cultural politics," sometimes appear atheoretical and uninteresting from the perspective of the new sociologists' debates. By writing an alternative text that celebrates at least some of the other voices, I hope to legitimate another set of questions and another way of conceptualizing the "theoretically interesting."

Culture and Power

I have used culture and power as two central organizing concepts for the way in which I am thinking in this book. I have purposely chosen not to articulate a theoretical framework in any more detail than this pair of central rubrics. I am suspicious of the masculinist and totalitarian implications of attempts to formulate programs and schools. There are already enough of them in STS, and there is no need for more. Instead, I have chosen the terms *culture* and *power* because they are widely used and accepted concepts among those researchers who have been advocating critical social studies beyond the "epistemological radicalism" of the new sociologists of science. The terms *culture* and *power* also emerged as a satisfactory common ground for a group of anthropologists who participated in panels together at some of the annual meetings of the Society for the Social Studies of Science.

My own use of the terms *culture* and *power* is rooted in my training and professional experience as an anthropologist. I do not assume that the terms represent the only valid frames for examining the world. In fact, cultural anthropologists are, like other social scientists, notorious for disagreeing among themselves. Some may think that disagreements are an indication of low scientific credibility, but debate and dissent can also be seen as an essential part of a democratic research community. Thus, my choice of central organizing concepts is contestable. I have a particular attraction to questions of cultural and power because research in that area has been underrepresented in STS, as has been research by cultural anthropologists. There are several conflicting uses of *culture, power,* and *cultural politics,* so it may be helpful to specify how I will be using the words in this book.

Culture and Society

Regarding *culture,* like most cultural anthropologists I do not restrict the scope of the word to the intangible area of ideas, religion, arts, literature, language, films, oral traditions, and, in short, the intellectual life of a people. Rather, the concept of culture in an anthropological sense has more to do with the total knowledge and way of life of a group of people: both conscious and unconscious, implicit and explicit. It is everything that a group of people has learned, and therefore it is best defined in contrast not to society but instead to nature, to a people's physical environment (although understandings of nature are themselves culturally constructed) and to their biological inheritance (which is mediated through constructions of biological scientists).

Thus, culture includes not only the intangible beliefs of a people but also the domain of social action: rituals, work, trade, political institutions, family and kinship, and so on. Sometimes the idea of the social is subdivided into social structure and social relations. By social structure I will mean the general ways in which social groups are organized in a society: by gender, age, class, ethnicity, race, occupation, institutional organization, networks, and so on. By social relations I will refer to the interactions among people and/or groups as they negotiate their way through institutions and social settings: their actions, strategies, motivations, roles, and interests.

My use of the culture concept is also recursive, cross-cutting, and flexible. There are all sorts of intersecting and nested cultures: transcontinental, continental, national, urban, regional, occupational, class, organizational, neighborhood, ethnic, gender, and so on. Some occupational cultures, such as physicists, have many similarities even in different national cultures, whereas other cultures are bound to local settings. Some cultures can be thought of as contained within others the way that regional cultures are contained within a national culture. Some cultures are more or less coterminous with societies and states, whereas others extend in significant ways beyond national borders (e.g., American Hollywood culture, French intellectual culture, Chinese immigrant neighborhoods). Cultures also change, so it is important to situate analyses during a given period in a society's or an organization's historical development. New cultures emerge all the time, especially around new knowledges and technologies but also in the borderlands among societies, in the diasporas of immigrant groups, and in the transnational networks of international communities. It is therefore unlikely that the world is becoming "more homogeneous": there are both centripetal and centrifugal tendencies, merging cultures and emergent cultures.

Culture cannot be reduced to a catalog of values, traits, ideas, behaviors, and institutions. One has to go inside a culture, talk to people, and find out how they see their world and how commonalities cut across different individuals and groups. Anthropologists refer to these commonalities as values, patterns, structures, or themes. Although at one point anthropologists assumed that common structures were more or less everywhere in the culture, today they recognize that there are many areas of difference and even conflict. Cultures are therefore dynamic and contested, even if there are areas of "doxa" that lie outside the contested domains of orthodoxies and heterodoxies.

A cultural perspective implies studying science and technology from the point of view of different groups of people. This perspective assumes that there will be differences of understandings, opinions, and evaluations. Science and technology may mean one thing to nuclear physicists and molecular biologists, but it may mean something completely different to a working-class neighborhood in Britain, a team of acupuncture researchers in China, a New Age commune in California, or a community of peasants in Peru. Every group has its own version of what they see as knowledge and technology. Thus, at certain points in this book, I will be using the words *science* and *technology* very loosely, more in the sense of knowledge and human-made artifacts.

To understand the question of meaning—"What does X mean to Y?"—anthropologists listen: we ask people, read what they write, or watch them. At the same time we compare what they say to the positions of other people, either within or outside the culture. Cultural analysis is therefore a very contrastive enterprise. To borrow the example of the "tribe" of anthropologist Sharon Traweek, it is not really possible to understand the culture of American high-energy physicists until one has compared them with the sibling culture of high-energy physicists in Japan . . . or, to get closer to an area of my own research, with the reconstructions of physics by Brazilian Spiritists.

I will use the term *society* in a specific sense as well. When a culture corresponds to a large group of people who occupy a territory and have institutions for provisioning and reproducing themselves, then the culture can also be said to be a society. A society does not necessarily have an independent government, as is demonstrated in the case of many ethnic minority societies. For example, the government of Canada has spent years attempting to develop a definition of Quebec as a separate society, even though Quebec does not have an entirely independent state. Likewise, the *same* culture can be split into more than one state, as in North and South Korea, which soon evolves into two societies. Finally, even when a culture coincides with a society, it can extend beyond the boundaries of a society, as occurs when immigrant groups form ethnic enclaves and neighborhoods in their new host societies.

Power and Politics

So far the discussion has focused on structural aspects of the analytical framework, on the concepts of culture and society. Yet, to understand the social world one must also understand how people act and navigate through their worlds. As people move through their social world, their actions have consequences. I am most concerned here with the consequences that have to do with equity and, therefore, with the issue of power.

By the terms *power* and *politics* I refer to a wide range of relationships that are not limited to government. To paraphrase the classical definition of the sociologist Max Weber in *Economy and Society*, power is the ability of people or groups to get what they want, even when other people or groups want something else. The simplest case is a leader who gives an order and expects subordinates to carry out the order. The subordinates may be motivated by fear, admiration, persuasion, material gain, habit, or any number of reasons. Power operates best when it is accompanied by a belief in the legitimacy of the people or groups who exercise it; in that case they can be said to have authority.

Weber's definition of power is considerably more general than the conventional idea of controlling a political office; however, today many social scientists prefer an even more general understanding of power that, in one version or another, owes some debt to the French historian and social theorist Michel Foucault. Foucault altered the theorization on power in a number of ways, but the one with which I am concerned here involves the ways in which power is embedded throughout society in a myriad of everyday practices. He used metaphors such as capillary and microphysics to signal this aspect of power. More than a question of who controls an organization or who has the ability to make successful orders, power involves social practices that have differential effects on individuals and groups. In practical terms, the differential effects with which I will be most concerned here are questions of equity. I will consequently use and understand the term *power* in a specific sense to involve any cultural or social relationship viewed from the perspective of differential effects on equity. Equity involves differential access of groups and individuals to economic resources, political rights, social status, and opportunities in general.

Because any social action shifts social structure or relations in some way, and because almost any shift affects some group or social category adversely or positively, then almost every social action has a power dimension to it. To understand what kinds of groups are relevant, one needs to enter the local culture and find out which categories are important. Thus, a cultural perspective is crucial to the analysis of power. In capitalist societies power and equity issues often play themselves out around access to resources and other aspects of the politics of economic relationships. Thus, by a culture-and-power perspective I am assuming that in capitalist societies (and perhaps other ones as well) relationships of political economy will be crucial.

Power enters into the making of science and technology in a number of ways. First, there is the question of which ideas and artifacts become widely accepted in the society and which ones pass into obscurity. In other words, at the level of production power shapes what science and technology gets produced and distributed. Second, as science and technology work their way through society, they have various differential effects on different social groups. To borrow an example from the political theorist Langdon Winner (1986), bridges that were built in Long Island so that buses from the inner city could not pass under them carried a politics of white, middle-class racism. I follow Winner in arguing that "artifacts have politics" (as do ideas). Thus, the technical *is* the political, just as the scientific *is* the ideological. Or, to put the same idea in a different way, there is always a political-ideological dimension to the technical-scientific.

I see the culture-and-power perspective as a necessary combination. Without a cultural perspective the analysis of power tends to be limited to a few social categories or groups that appear important to the researcher (usually class, race, gender, and nationality/ethnicity). Yet, the narrower the scope of categories, the less one is able to see the power dimension and the more one's own analysis contributes to the legitimation of power relationships that may be invisible to all but the disempowered. Conversely, the larger the number of salient social categories available for analysis, the greater the number of opportunities for understanding the effects of power. Furthermore, because questions of power and equity are in the eyes of the beholder, social scientists also need a cultural perspective in order not to impose their own views of equity on the social groups in question.

A cultural perspective is also crucial for getting at the complexities of the way power operates in society. Individuals always have a social position that locates them in some ways "up" and in some ways "down," depending on the situation. Take, for the sake of simplicity, the classic triad of race, class, and gender in many Western countries. The relationships between white, upper-class men and black, lower-class women may be interesting, but there are a number of other ambiguous relationships, such as those between white women and black men. Furthermore, as people move across roles and through life, their social position changes. In the Americas a white, upper-class boy may live part of his youth in a world constructed by a lower-class woman of color. That is one way in which a cultural perspective can complicate and inform an analysis of power.

Given these complexities, there is ambiguity in terms such as *privileged, dominant, historically oppressed, disempowered,* or—to use the legal terminology—*underrepresented ethnic groups.* Although I use those words in the pages that follow, their limitations should be clarified at the outset. A cultural perspective suggests that one needs to examine such labels by seeing how people view themselves. The labels can be useful starting points that flag all-too-easily forgotten social structures of neo-colonialism, racism, patriarchy, heterosexism, and class domination. However, the labels should also be seen as only pointers to changing social locations in specific cases. The marginalized have a habit of moving out of the margins, and the disempowered of making power for themselves. They are not passive victims. To develop this point, I explore in some detail the idea of reconstruction or construction by excluded and lay groups.

Furthermore, marginal status is linked as much to what one says and thinks as it is to the physical markers of one's body or one's address in the general social structure. My interviews with parapsychologists show how educated, white males in the North American academy can suffer from processes of marginalization similar to those experienced by radical feminist, antiracist scholars (Hess 1992a). Generational changes further complicate the picture. In the United States older women and people of color have fought hard for the gains that a younger and sometimes more conservative generation takes for granted. It is useful to keep straight the interrelated difference between social stand and social position, that is, between "where one stands" and "where one sits."

Another area where word choice is very complicated involves the terms for ethnic and racial groups, with particular reference to groups within the United States. The standard terminology used in the 1990s in the United States—white/European descent, black/African American, Hispanic/Latino/a, Asian American, and Native American—reveals category mixing that includes skin color, descent, and language. Although I frequently use those basic categories, again it is with a sense of their limitations. In situations of coalition politics, differences within each category are often more salient. Furthermore, the categories are changing. The idea of a European American would have been bizarre in the nineteenth century, when NINA (No Irish Need Apply) signs hung on storefronts and cultural differences among European populations were often biologized in racist discourse. It is also unlikely that by the year 2050, when the white/European population of the United States is projected to be in the "minority," the same categories will be in operation. The phenomena of overlap (e.g., Latins who are "white") and miscegenation will likely lead to such radical changes in the classification system that the very concepts of ethnic majority and minority are likely to undergo radical changes as well.

In the chapters that follow, I focus on cultural politics in the study of science and technology. In a huge field of research I have limited my examples to some well-known cases in history and some somewhat less-

er-known cases in anthropology. Not all the chapters explore the culture-power connection in the same way or at the same level. Some chapters focus more on culture, others more on power. Yet, the two rubrics
are the central concepts behind my effort to bring together, in one
place, a body of research and a set of arguments that have to do with
science and technology from the perspective of cultural differences and
social justice.

Overview

Chapter 2 amplifies the theoretical basis for the book by discussing in
more detail cultural construction and reconstruction. Beginning with
anthropological theories of totemism and mythology (*bricolage*), I show
how culture and power can contribute to taking social studies of science
and technology beyond social constructivism. The chapter also provides some examples and case studies in order to make the concepts
less abstract and more meaningful. Chapter 3 applies the concepts
developed in chapter 2 to a critical survey of the standard histories of
modern science and technology. Focusing on narratives of the origin of
Western science, this chapter shows ellipses in the narratives where
non-Western events have been written out of the history. In addition, the
chapter elucidates some of the theories of the scientific revolution in
terms of class, gender, and colonialism. Rather than provide a survey of
the history of the scientific revolution, the chapter provides a framework
for developing critical rereadings of history. Chapter 4 develops the idea
of temporal cultures and explores how the scientific theories of a given
time period can be seen as coherent with the general culture of the period. The chapter also ends with a discussion of contemporary science in
the emergent global society and postmodern culture, with a cautionary
tale on what I call the "boomerang" of technototemism.

Whereas the third and fourth chapters explore culture-and-power in
theories across time, chapter 5 does so with social relations and structures across space. That chapter shows how communication processes
and social organization vary across cultures, thus helping to put into
question the assumption that modern science and technology are everywhere the same. Chapter 6 further expands the scope of the cultural
analysis by showing how nonexpert groups reconstruct science and
technology, thus challenging the authority of the experts as well as the
politics embedded in their theories or artifacts. The study of how non-

professionals construct science and technology is extended in chapter 7, which explores indigenous knowledge and non-Western medicine, including their impact on and coexistence with cosmopolitan science and technology. Chapter 8 examines indigenous perspectives on cosmopolitan technology and development. In an otherwise very sad story of the relations between powerful and relatively powerless societies, the chapter also points to resistance struggles and grassroots development projects as alternative models for relations between indigenous and cosmopolitan societies. Finally, the conclusion considers the question of multiculturalism, contemporary society, and educational reform.

2 | The Cultural Construction of Science and Technology

Thinking culturally implies beginning with the question of local meaning: What does X mean to Y? For example, what does this computer mean to this group of experts as opposed to that group of experts, or as opposed to that group of nonexperts? To answer the question one maps out technical categories or distinctions. At any given time the various groups within a field of research are producing different theories, observations, machines, methods, experiments, and so on. In some cases those differences are merely the result of asking different research questions, and they represent complementarities rather than contradictions. In other cases differences take the form of contradictions, and controversies and debates emerge. Whether the differences arise from complementarities or from contradictions, the differences have meaning. Those meanings may be explicit—consciously understood in the minds of the people in question—or implicit, that is, tacit in their actions and linguistic codes.

At the most internal level the meaning of technical differences can be interpreted by referring to other technical distinctions. This procedure for generating meaning is generally how textbooks work: a new concept is explained by building on previous ones. However, often the technical differences have some relationship to social differences as well. Sometimes the social differences are local and internal to the communities in question—such as two theories that map onto competing laboratories or research networks—but sometimes the social differences correspond to divisions in the broader social structure, such as class, gender, sexuality, ethnicity, and nationality. As the anthropologist Emily Martin has shown in *The Woman in the Body* and in numerous articles, often even textbooks suggest broader social meanings in the metaphors they use to explain technical ideas and in their cartoons and diagrams.

In contrast to this cultural or meaning-oriented approach, more sociologically oriented analyses tend to work in terms of explanations and variables. They show how a given set of social factors, X, can be operationalized as a variable that influences the final shape of a scientific knowledge or technology, Y. In a formula social constructivism looks at the technical in terms of a logic of explanation: X (social factor) influences Y (scientific idea or technological artifact) and therefore explains how Y got to be the way it is. Social constructivist analyses therefore often provide narratives of how facts or machines came to be widely established. The social factors (X) and technical content (Y) do not necessarily have to correspond to local categories; instead, they represent categories that the researcher finds relevant for making the analysis.

In contrast, a cultural or meaning-oriented approach to construction is interpretive (see figure 2.1). In other words, the object of interpretation Y is broken down into salient distinctions in the culture (Y1, Y2, Y3, etc.) and is then interpreted in terms of how the divisions within Y translate into other salient cultural categories, such as the divisions within X (X1, X2, X3, etc.). By contrasting social constructivism to cultural constructivism, I do not wish to imply that one approach is right and the other is wrong. It would be better to think of the two as complementary, as different ways of approaching the same phenomena. Furthermore, no one social scientist, and no single social scientific study, can be placed clearly in one category or another. Instead, it is more likely for the researcher to use both types of analysis at different points in a discussion or essay. Thus, the two forms of analysis might be seen as the two ends of a continuum. Even in this book the present chapter tends toward the cultural, or "anthropological," side of the continuum, whereas other chapters will focus more on explanations of social influence, and in this sense they might be considered more "sociological."

FIGURE 2.1
Social Versus Cultural Construction
Two Hypothetical Examples

Social Construction: Causal Relationships Among Variables

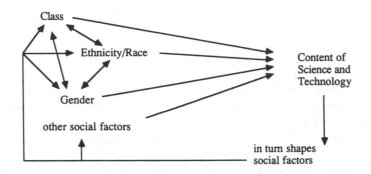

Cultural Construction: Mappings of Categories of Meaning

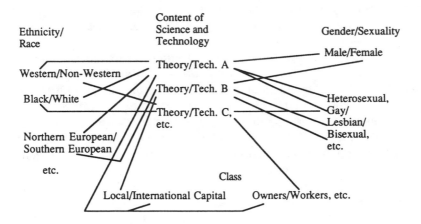

Technototemism

I refer to the coproduction of technical and social difference as technototemism, a concept that will constitute the starting point for my version
of a culture-power perspective. Totemism is the process by which social
groups achieve coherence and distinctiveness by being identified with
natural phenomena. For example, a member of a traditional indigenous
society might belong to the bear or owl clan. Today, sports teams are frequently marked by natural totems, for example, the St. Louis Cardinals
or the Chicago Bears. Of course, some cultural objects also have become
totems, such as the Cincinnati Red Sox. Likewise, some of the totems
have become extremely controversial, such as the Washington Red Skins
or Cleveland Indians. Part of the controversy—and the offensiveness of
the practice—stems from the fact that by naming one group of people (a
sports team) after another group of people (Native Americans), the second group of people is put into a natural/material category along with
animals and socks rather than into the human category of sports teams.
(That naturalization extends to other kinds of stereotypical behavior
that the mascots may perform as well.)

Offensive as it sometimes can be in today's world, totemism in a general sense is a widespread part of human societies. As the anthropologist
Marshall Sahlins has shown in *Culture and Practical Reason*, a great deal
of consumer culture operates according to totemic relationships. Clothing, food, cars, and so on are all categorized into a myriad of divisions
that allow people to make distinctions among themselves through their
objects. If a white, male American drinks Coors beer, drives a Ford pick-
up truck with an NRA bumper sticker, and wears checked flannel shirts,
he is simultaneously saying something about himself as someone who
does not drink organic soy milk, ride public transportation, and wear T-
shirts with feminist or antiracist drawings on them. In this way his social
identity—the social category to which he belongs—becomes solidified
or clarified through his location in a set of material codes. Likewise,
those codes become solidified and clarified by having a social location
or address.

Advertisers understand very well the way categories of consumer
products are interwoven with categories of gender, ethnicity, class, and
regional identity, and they base many of their campaigns on establishing
those differences for consumers. If they bombard consumers with adver-
tising such as "This is not your father's Oldsmobile," sooner or later the

Olds-mobile/old-generation totemic linkage will be replaced with another one. In other words, totemic relationships are not naturally given; they are constructed through the actions of one social group on another. Those actions are not innocent of power. Large corporations spend fortunes every day to get consumers to make the proper associations in their minds. As a consumer/viewer, I may write the television station and ask it to run another advertisement that creates some other totemic linkage with the advertised product, but I am not as likely to be heard as someone who owns major shares in the company or pays thousands of dollars per advertisement. In short, totemic relationships involve more than meaning in the sense of semantic difference; they also involve power.

A fundamental attribute of totemism is that social categories and natural/technical categories are coconstituted as systems of distinction. In other words, the social and natural/technical become meaningful by being mapped onto each other or translated into each other. This is not to say that without brand names it would be impossible to distinguish a redneck from a radical activist; other codes, such as repertoires of behavior patterns or conscious beliefs, could also be used. However, just as part of what it means to be a member of the bear or owl clan depends on the differences attributed to bears and owls, so part of the identity of being a redneck or a radical depends on the meanings attributed to the different consumer products that one social category uses in opposition to another. In its simplest form, a division of social categories is brought to life via a division of natural categories (or, in our technological culture, of material or technical categories), and vice-versa.

It is also possible to speak of scientific totemism or the totemism of scientific ideas. As Sahlins wrote, "[O]ur science may be the highest form of totemism" (1976:53). His example drew on evolutionary biology and a very insightful quote about Darwin by his contemporary Marx. In a letter to Engels in 1862, Marx wrote:

> It is remarkable that Darwin recognizes among brutes and plants his English society with its division of labor, competition, opening up of new markets, "inventions," and Malthusian "struggle for existence." . . . [W]ith Darwin the animal kingdom figures as bourgeois society. (Sahlins 53)

Through natural selection Darwin injected mobility and dynamism into the scientific models of the natural world: plants and animals may evolve, become extinct, or exchange positions over time, just as capitalist firms may rise and fall. As Sahlins goes on to explain, Engels then noted that once the transfer of new social relations onto nature had been completed, nature could be transferred back onto society as a means of legitimating an existing social order as the natural order. I call this the "boomerang" of technototemism; others have used the words *reification* or *mystification* to describe related processes. In other words, because relations were such a way in nature, some people could therefore argue that those natural relations formed an ideal model for society as well. For example, Social Darwinism could be used to defend social injustice under the banner of the survival of the fittest.

Sahlins restricts his discussion of scientific totemism to the case of Darwinian evolutionary theory. That theory can be read as almost a literal example of totemism, because relations among plant and animal species serve as mirrors through which relations among human social groups can be articulated. I will use the concept of technototemism in a somewhat broader way to refer to any case in which divisions among social groups become identified with technical/natural distinctions, that is, distinctions among ideas, theories, schools, technologies, natural facts, or methods. The STS literature is full of case studies that show how social categories correspond to technical ones. Given the general topic of the book, multiculturalism, I have chosen in this chapter to focus on only a few of the possible social categories: class, gender, race, and nationality.

The Inscription of Class, Race, and Gender in Technototems

To give some life to the idea of technototemism, or totemic relationships in science and technology, I present a few case studies that focus on how class, race, and gender become articulated or translatable through technoscientific distinctions. There are an indefinite number of social relationships, both macro and micro, that are reproduced in a field of technical differences among theories, facts, methods, devices, and so on. I begin with a discussion of class interests, not because I want to privilege class, but because a specific debate emerged over class in the STS literature, and that debate is of general interest to any discussion of technototemism.

Class Interests and the Imputation Problem

A now classic example of scientific totemism in the STS literature involves constructions of statistics in early twentieth-century Britain. As the historian Donald MacKenzie has shown, a debate emerged between the advocates of Karl Pearson's correlation statistic r_T and those who supported the alternative Q statistic of his former student George Yule. Pearson's statistic assumed that a division of data into two categories (such as tall and short people) could be modeled as an arbitrary division in a normal distribution, that is, a bell-shaped curve. For example, the height of people in a large classroom tends to form a bell curve, with the most people having the mean height (say, six feet in a given sample). The farther one goes in the distribution from the mean height, fewer and fewer people will have taller or shorter heights. Pearson argued for applying a powerful statistic—one that assumed an underlying, bell-shaped normal distribution—to cases of nominal data, that is, data that are divided into two categories such as tall and short.

In contrast, Yule argued for a statistic that did not require the assumption of an underlying normal distribution. For example, in the case of a population that has been exposed to smallpox, there are only two alternatives: they are either dead or alive. This type of data is now known as "nominal data," and Yule argued that it did not make sense to assume that there was an underlying normal distribution for nominal data. As Yule wrote, "All those who have died of smallpox are equally dead" (Barnes and MacKenzie 62). Today, both types of statistics are used, and the controversy has been largely resolved by a pluralistic view that denies unique validity to any one statistic. However, at the time things were not so clear, and a debate emerged.

The debate appears to be a purely technical one having to do with the kind of assumptions one wishes to make about underlying distributions. However, as MacKenzie shows, the technical debate is simultaneously a social conflict. Pearson's supporters were tied to the biometric and eugenic laboratories of University College London, and Pearson was the leader of that group. Pearson's defense of the correlational statistic that assumed underlying population distributions was closely linked to his concerns with heredity and social change through eugenics. As MacKenzie and his colleage Barry Barnes note, Pearson's work was "a project aimed at providing a scientific basis for eugenic policy and ideology," that is, a political program that supported selective human breeding in order to improve the population (Barnes and MacKenzie 59–60). Yule, in contrast, had a following among other researchers, primarily in the Royal Statistical Society. He was more concerned with public health applications such as vaccination, and to measure success and failure in that area one needed only a simpler statistic that measures the correlation between vaccinated and nonvaccinated versus dead and alive. It is therefore not surprising that "Yule moved away from the biometric laboratory where Pearson had taught him into the more conservative context provided by the Royal Statistical Society" (61).

MacKenzie therefore shows how the distinction between two statistics tends to line up with a social distinction. At an interpersonal level it is a distinction between two people, but Pearson and Yule also drew supporters from two networks of statisticians, respectively from the University College London and the Royal Statistical Society. Behind the two networks were clashing political views that MacKenzie argues can in a general way be associated with conflicts between the professional class and the established upper classes. Pearson was a member of the emerging middle or professional class, for whom eugenics was associated with professional power and differentiation from the working classes. Pearson's political views were close to Fabian socialism, a paternalistic form of socialism that would benefit the professional classes through the creation of a welfare state. In contrast, Yule was a conservative aristocrat who had no use for eugenic social programs. The link between the scientific and the social is a multiple one: clashing individuals, rival networks, and conflicting classes. In this sense the two statistics serve as technototems.

Barnes and MacKenzie viewed the Yule/Pearson debate through the lens of interests, particularly class interests. Their analysis became the focus of criticisms that raised a number of objections to interest-based explanations of science and technology. Among the criticisms were questions about the credibility of attempts to impute class interests to Yule and Pearson. There is very limited evidence that the statisticians were consciously developing the statistics to serve their class interests, and even if they were there would have been many other motivations or causes for their competing approaches. Thus, there is a problem with assigning—or imputing—cause-and-effect relationships between the macrosocial class configurations and the microsocial politics of conflicting networks and statistical measures.

However, for the project of critiquing the cultural politics of the statistics, it is not necessary to develop a causal or explanatory analysis of why the statisticians came to develop their statistics. Rather, the approach that I have been outlining sidesteps the problem by interpreting the cultural meaning that is implicit in the rival statistics. A first step would be to examine how the actors understood their debate and its relationship to competing networks and classes. (That could be accomplished either through documents or, in contemporary cases, through retrospective interviews and fieldwork.) As I read the debate, I find myself asking the anthropological question, "Well, what did Pearson and Yule think about the general social significance of their work, including its relationship to class interests?" Apparently there is little in the way of historical documentation to answer the question, and it is impossible to resort to interviews or ethnographic methods because both statisticians are now equally dead. In the absence of more evidence, it is better to limit the relationship between the statisticians and class cultures to one of correspondence rather than causality. In other words, rather than say that class interests explain the different positions in the statistical controversy, it is sufficient to say that the problems and social programs that seemed relevant in the two class cultures were similar to the problems and social programs that seemed relevant in the two statistical networks. Thus, from the existing evidence it seems reasonable to posit a totemic relationship between class divisions and statistical measures. For the purposes of cultural critique it is not necessary to move to an explanation (class causes statistics); it is enough to demonstrate that there is some evidence that the two levels of difference coincide.

The problem with using interests as a causal explanation of scientific theories was one reason why many STS researchers rejected the interest-type of analysis pioneered by Edinburgh school researchers such as Barnes and MacKenzie. Unfortunately, the wholesale rejection of their research agenda may have meant losing a valuable tool for the cultural critique of science and technology. Although I would agree with critics who find a problem with assigning a causal relationship between, say, social classes and scientific theories, the relationship may still exist in terms of similarities or parallels across cultural domains. In other words, one may be able to establish a structure across domains even if a causal relationship cannot be demonstrated. The "meaningfulness" of the structure can be implicit or it can be explicitly recognized by the actors involved.

There is also a problem in the limited focus of explanation at play in the debate. It may indeed be problematic to say that the "cause" of the difference in the statistics and the debate over the statistics is, at least in a simple way, class interests. One can answer the problem by saying that class may be reasonably postulated as at least one important shaping factor in a complex web of causes. Nevertheless, the entire debate would still focus on the front-end of causal relationships, that is, relationships that explain the existence of the correspondence. Whatever one decides "causes" the existence of technototemic relationships in the first place, those relationships may in turn contribute to the reproduction of technosocioconfigurations long after they have left their context of production. In other words, technototemic relationships are not only effects of social conflicts or social forces but also causes in the ongoing legitimation and reproduction of a social and intellectual order. This "seamless web" is better demonstrated in some of the studies of the technototems of gender, race, and ethnicity.

Technototems of Gender and Race I: Cells

Biology, medicine, and psychiatry have proven to be a key site for social studies of knowledge construction in terms of gender and race. This first section will consider how controversies in cell biology often appear first as intellectual debates but can be unpacked to show that they are translatable into structures of race and gender difference. My discussion follows the general comparative argument developed by the Biology and Gender Study Group (1989) of Swarthmore College, which involved the collaboration of a group of mostly women students with the historian Scott Gilbert.

In the essay "Cellular Politics" Gilbert shows that in the 1930s the African American marine biologist and embryologist Ernest Everett Just argued in favor of the role of the cytoplasm over the nucleus in the development of the organism. As Just summarized his position, "In contrast to the conception of the gene-theory, my fundamental thesis is that the genes act by removing stuff from the cytoplasm" (293). Scott demonstrates how Just's support of the role of the cytoplasm was related to his identification with embryologists, who at that time were disputing with geneticists the importance of the nucleus in the development process. Embryologists favored a larger role for the cytoplasm, and geneticists favored a larger role for the nucleus. At stake was the relative importance of two disciplines, and the potential implications of the debate included access to institutional support. In *Beyond the Gene* the historian Jan Sapp discusses this larger history of the controversy between nucleocentric and cytoplasmic approaches to inheritance.

Both Sapp and Gilbert have analyzed some of the ways in which this controversy maps onto wider macrosociological codes. One example involved the way structures of the cell were used as metaphors for society as a whole, with the nucleus representing the elite and powerful groups and the cytoplasm the rank-and-file or masses. Gilbert argues that Just's attention to the importance of the margins of the cell rather than the nucleus therefore corresponded to other relationships between margins and center in American society. Just criticized the nucleus-centered approach to genetics as "a veritable decree of authoritarianism" (1936:292). Although Just was not the only biologist to criticize the political geography of the cell, questions of authoritarianism and marginalization had a special meaning for Just, who suffered from racism both within the scientific community and within the broader society. On this point it is interesting that he also attacked the "Theory of embryonic segregation"—a word that is highly charged in American English—and advocated instead a perspective that he labeled "differentiation" (1936:273–76).

So far the analysis is structurally similar to that used for statistics: a technical difference (two statistics/two parts of the cell) is shown to correspond to a microsociological difference (two networks or disciplines), which in turn maps onto macrosociological differences. However, ending the analysis at this point gives a false sense of closure. Instead, viewing technical difference as a cultural text to be interpreted rather than as a variable to be explained implies that the text can always be reread and the analysis is never really complete. One example, discussed by Sapp, involves the role of genetics in the emerging East/West conflict, in which Soviet scientists under T. D. Lysenko rejected Mendelian genetics. In this case the nucleocentric/cytoplasmic difference came to be aligned with national political differences in the emerging cold war. Another example involves the work of the geneticist Barbara McClintock for the case of cell structure. Like Just, she regarded herself primarily as a scientist whose theorizing took place within the intellectual and methodological confines of her discipline. Although she was a geneticist, she was very interested in questions of development, and she also worked with corn rather than bacteria, which most geneticists preferred. That background helped her, like Just, to see complexity in the relationships among genes, the cell, and development. McClintock also suffered severe gender discrimination similar to the racial discrimination suffered by Just; both ended up in relatively marginal posts rather than as professors in prestigious departments. McClintock eventually received the Nobel Prize for her work in genetics, but recognition came only late in life.

The parallels between McClintock and Just go beyond their life histories as mid-twentieth century American biologists who suffered discrimination. As Just had argued in favor of a greater role for the cytoplasm, so McClintock opposed what came to be known as the master molecule approach. That approach or theory was associated with Watson and Crick's "central dogma," in which the DNA produces the RNA that produces protein. As the STS researcher Evelyn Fox Keller describes, "In locating the seat of genetic control in a single molecule, [the master molecule theory] posits a structure of genetic organization that is essentially hierarchical, often illustrated in textbooks by organizational charts like those of corporate structures" (1985:170–71). In contrast, McClintock's less hierarchical theory showed that genes could be rearranged and that therefore they could respond to external signals from the cell or the environment. Two totems, again, were opposed: McClintock's transposable genes versus the geneticists' master molecule.

Was McClintock's theory related to her position as a woman in a field dominated by men? Keller is careful in answering the question. McClintock was not a feminist and, like many women scientists, adhered to the ideal of a gender-free science. However, gender may still be part of the story at the level of politics and cultural logics. In feminist circles Watson and Crick are known for antifeminist attitudes ("The best home for a feminist is in another person's lab"), for using without permission or acknowledgment crucial data of the crystallographer Rosalind Franklin, and even for carelessly letting their softball land too often in McClintock's cornfields at Cold Spring Harbor. It should therefore be no surprise that feminist science studies researchers have questioned the gender implications of the master molecule theory's simplistic assignment of activity to the nucleus/genes and passivity to the cytoplasm/cell environment. Thus, whatever her personal motivations, McClintock helped to put into question a central dogma of her field that some have argued reflects a masculinist bias. (Of course, in the context of Just the dogma is not restricted to a masculinist bias; the term *master* takes on entirely new meanings in that context.)

These studies point to how one cultural order (models of the cell) translates into other ones (codes of race and gender), even when the scientists themselves may not be entirely explicit about or conscious of the translations at work. One way of making analyses of this type more robust is to follow the metaphors that scientists use in their theories and descriptions. Those metaphors provide evidence of translations among domains that a structural analysis can only suggest. For example, one way in which the gendered aspect of the relationship between nucleus and cytoplasm has been established has been through metaphors used in association with the fertilization process. Because the sperm is seen as contributing only to the nucleus of the zygote, the cytoplasm remains a feminine space. In other words, the gendered sperm/egg relationship translates into the nucleus/cytoplasm relationship.

The Biology and Gender Study Group has worked out how masculine narratives of heroic conquest were woven into biological descriptions of insemination as the work of a powerful sperm that competes with other sperm as it fights it way to the passive egg and penetrates through its wall. The activity of the sperm is then transferred inside the cell to become the active nucleus, whereas the passivity of the egg becomes the passivity of the cytoplasm. Subsequent biological research has undermined those narratives of insemination substantially. That research shows how the egg plays a much more active role in attracting the sperm to it, and that the tail of the sperm tends to jerk it from side to side rather than propel it onward and upward to the egg that it would conquer. However, as Emily Martin cautions, the active egg view of fertilization may undermine the implicit sexism of female passivity, but the new view may still inscribe its own antifemale or antifeminine gender biases (1991). She shows how the active egg models sometimes transform the passive female egg into the opposite, the aggressive and dangerous female that "captures and tethers" the helpless male sperm. In other words, the egg seems to have been transformed from virgin to witch or whore. The new construction therefore continues to borrow on general cultural stereotypes of gender in order to reproduce a logic in which the egg/sperm relationship recapitulates patriarchical views of feminine/masculine relationships.

From these examples it becomes possible to see basic cellular structures as a complex text that can be read via general social codes of gender and race. The master molecule theory of the central dogma ascribes agency to the chromosomes and genes of the nucleus, in contrast to passivity for the proteins, cytoplasm, and environment (nature) beyond the cell. That story may turn out to be an acceptable map of the world, or future research may discredit the story as too simplistic. In either case, the ways the story has been told convey a social message in which, to exaggerate, the white boys are preaching to everyone else the dogma of who is the master of the house of cell, science, and society.

I have chosen these interrelated examples because they show a way of thinking about gender, race, and the content of science that avoids the imputation problem. They do so by focusing on the social/technical connection as one of translation rather than causation. In other words, it is not necessary to inspect McClintock's conscious gender politics or Just's explicit race politics and argue that they intentionally built their theories to reflect their position as a woman or as an African American man in the scientific community. It is not even necessary to argue that their marginalized position in their scientific communities made it easier for them to be less invested in centralizing and hierarchizing narratives, or even, as in the case of Just, explicitly to suspect them. Instead, it is only necessary to follow out the symbolic connections among nucleus/cytoplasm, sperm/egg, white/black, and male/female as cultural orders or codes in a society. Thus, the controversies over cell structure serve as points of translation between a social hierarchy and a cellular one. The relationship is one of totemism, not necessarily one of causality. It is not a question of how variable A (marginality) shapes variable B (biological theory), but a question of how the structures and processes of social marginality and hierarchy are interwoven with those of biological theory.

These cases also avoid another type of imputation problem associated with the gendering or ethnicizing of methods. Some scholars have suggested that nonhegemonic groups—women, non-Westerners, Westerners of underrepresented groups, etc.—produce their science through different methods. Examples include the argument, for example, that women's methods are more qualitative or empathetic. One sometimes finds, especially in the popular literature, statements of the type, "Women or people of color have a more organic or empathetic approach to gathering knowledge about the world." However, the empirical research from science and technology studies—especially laboratory studies such as the work of the STS researcher Karin Knorr-Cetina in *The Manufacture of Knowledge*—suggests that there is no single scientific method and that scientists and engineers are quite opportunistic in their everyday use of methods.

Furthermore, feminist STS researchers such as Sandra Harding, Helen Longino, and Donna Haraway have cautioned against the dangers of positing a single feminist (or Native American or African American) method. An alternative, feminist/cultural approach would instead ask how methods of data gathering and interpretation can line up with gender or ethnic distinctions in specific circumstances. Women, for example, do not always bring more empathetic methods to a scientific speciality. Indeed, in some cases women and people of color have brought to fields of public debate and social scientific inquiry rigorous quantitative methods that help undermine longstanding popular perceptions. For example, in the 1890s the African American activist-scholar-journalist Ida B. Wells developed a quantitative analysis of lynchings, based on documents written by whites, to debunk popular stereotypes that lynchings were associated mostly with black men who had raped white women. Her methods were a rhetorically important choice for the time period and issue.

In some cases universalizing or essentializing claims about methods can be politically useful rhetorical devices, but in other cases they may have the unintended consequence of perpetuating beliefs that women, people of color, and other historically excluded groups always think more empathetically/emotionally and less rationally. Beliefs of that sort in turn can legitimate exclusionary social practices that prevent those groups from making it into the classrooms and laboratories of technological production. As a result I have focused here on the content of what gets produced rather than the methods by which that content is produced. Although methods can also line up with distinctions of race and gender, the analysis should be grounded in case studies to avoid making universal claims about the question of difference.

Technototems of Gender and Race II: Intelligence

A second major area in which the structures of racial and gender hierarchies are played out in scientific theories involves psychology, psychiatry, physical anthropology, and other fields that work with human difference. This section will focus on controversies over intelligence, which throughout the twentieth century have been a site for the translation of racist and sexist ideas into scientific discourse. The controversies over intelligence are also interesting because they involve a translation of theories across the French/Anglo-Saxon cultural divide.

The construction of mental abilities and race has a long history, from debates over skull sizes and the ill effects of racial mixing in the nineteenth century to debates over IQ tests today. The word *miscegention* was originally a pejorative term for the mingling of the races, which would lead to the destruction of civilization through its replacement by a world of "mulattoes." Originally a term of disparagement for persons of mixed ancestry, *mulattoes,* like their etymological relative *mules,* were considered degenerate hybrids that threatened the hierarchy of races celebrating the civilized, white, and empowered over the "primitive," dark, and enslaved. By the twentieth century most of those theories had fallen out of favor, but in their place racist accounts of mental abilities emerged via the new methods of psychological testing.

As developed by Alfred Binet in France in 1905, an intelligence test measured the mental age of students in order to find children who could benefit from remedial intervention. In *Not in Our Genes* the scientists R. C. Lewontin, Steven Rose, and Leon Kamin describe how Binet emphasized intelligence as a socially shaped ability. Although Binet recognized obvious biological aspects of intelligence such as retardation, he carefully drew the limits for biological explanations of intelligence. For example, he explained the difference between upper-class Belgian scores and Parisian scores as due to classroom sizes, school quality, and home environment. Binet's overall sociocultural approach to intelligence was quite at odds with approaches that soon achieved prominence in the English-speaking countries.

In Britain and the United States intelligence testing fell into the hands of the eugenicists, including Pearson's student Cyril Burt in Britain. In the United States the Binet test was transformed into the Stanford-Binet intelligence test, which used the intelligence quotient (IQ)—that is, the ratio between the mental and the chronological age that had been developed by the German psychologist William Stern in 1912. (Clearly, this was the kind of variable for which Pearson's r would be more appropriate than Yule's Q. In contrast, Yule's Q would have worked better if one thought of intelligence in terms of "normal" and "retarded," which was all that Binet really required for his approach to remedial education.) Eugenicists developed the view that intelligence was inherited and immutable, and from the beginning the test was used to separate out groups by race or class for educational "Tracking" (U.S.) or "streaming" (U.K.). The debate on racial differences in IQ has extended into the late twentieth century, notwithstanding multiple criticisms that include the exposure of Cyril Burt as a scientific fraud, the identification of cultural biases in test questions, and the hidden environmental variables ignored by the twin studies.

As totems, then, two definitions or theories become articulated along with two social positions. One side sees intelligence as a biological trait that can be quantified by testing and that varies across human racial groups. The other side sees the first position as racist and argues instead that intelligence tests measure school performance at most. The second side also argues that intelligence is a complex and multifaceted phenomenon that cannot readily be reduced to a test, that "races" are not genetically coherent, and that any attempt to link race to intelligence is unscientific.

To widen this cultural analysis of intelligence, it is interesting to compare the race/IQ debate with a theory about machine intelligence. Let us zoom forward a few decades to the middle of the twentieth century, when the mathematician Alan Turing developed the Turing test, which specified the conditions under which a machine could be said to think. Turing's work has been highly influential in today's debates about artificial intelligence, but it is also interesting to raise the question of his status as a gay man and the relevance of his sexual orientation to a theory that defines intelligence in machines. Although it would be hard to prove a causal relationship between his sexual orientation and his theories of mind and machines, that difficulty should not prevent an exploration of the implications of the connection from a cultural perspective.

Turing theorized a condition in which there would be a lack of difference between human and machine intelligence; in other words, he stated conditions under which a machine could be said to be equal to a human in intelligence. Whereas previously machines were regarded as incapable of intelligence, Turing postulated the conditions under which parity could be constituted. Turing also believed that homosexuality was not a pathology and that the persecution of homosexuals as biologically inferior was ridiculous. Given his beliefs about the equality of homosexuals and heterosexuals, it is interesting to raise the question of the extent to which the Turing test might also be read as a translation of the principle of heterosexual/homosexual equality into that of human/-machine equality.

Although this equation of cultural codes is only suggestive, the case of Turing's prosecution and punishment for homosexuality in the 1950s provides some very concrete information about how ideas of gender, sexuality, and intelligence were being mutually translated at mid century. At the time homosexuality was a crime, and when Turing was convicted he was treated with a hormonal therapy as an alternative to a prison term. The gender and intelligence issues come out in the choice of the hormonal therapy: doses of estrogen. Researchers during that period claimed that the hormonal treatment successfully reduced the sex drive, but they also suspected a side effect of inhibiting intellectual or learning ability. The belief that the female hormone could lower intelligence continued a longstanding medical tradition that constructed women's mental ability as biologically inferior to that of men. In the nineteenth century, neurologists warned women not to study too much lest they develop neuroses. Reason, and education, were the province of the white man.

Returning now to intelligence testing, in this context of medicalized discourses on homosexuality, female hormones, and intelligence, the IQ test may have actually represented a progressive move in terms of gender, even if it did not in terms of race. By the 1930s standard IQ tests had been developed that eliminated sex differences, which would have complicated the eugenicist project. Thus, there were contradictions even among those who believed that biological difference and intelligence were strongly related, and analyses need to be sophisticated enough to comprehend complexities of this sort.

In general, attempts to theorize women, gay men, or people of color as biologically inferior are part of a wider confluence of discriminatory social practices that continue to leave their mark in low enrollments of women and historically excluded ethnic groups in the technical disciplines. Western cultures have tended to situate women, people of color, and other Others as closer to nature. Located closer to nature, the white man's Others have been constructed as farther from reason and culture, and therefore in greater need of domestication, control, civilization, and the other accoutrements of the "white man's burden." Science—the locus of reason and culture, the province of the Western, the male, and the white—has all too readily answered that call in the past.

Technototems of National Culture

So far the discussion of technototemism has been limited to race, class, sexual orientation, and gender. It could easily be expanded to include the ways in which other markers of social difference are woven into scientific theories. Yet, there is another way in which social differences and technical differences map onto each other: within the largely elite, male communities of international science across national cultures.

Even for cultures that are in many ways very similar it is possible to locate dramatic differences in what is sometimes referred to as "national styles" of scientific thinking. For example, the French physicist and historian of science Pierre Duhem made a detailed argument that there was a tremendous gulf between English and continental physics that lasted at least until the late nineteenth century. Duhem characterized British physicists such as Faraday, Maxwell, and Thomson as concerned with building mechanical models or manipulating algebraic formulas. The English had a pragmatic, utilitarian attitude; consequently, they were not concerned with contradictions among formulas or the implausibility involved in using mechanical models for phenomena such as electricity. In contrast, Duhem characterized the continental style of physics and mathematics—the work of Ampère, Poisson, and Poincaré—as operating from first principles in the tradition of René Descartes. The "French mind" tended to work out a series of logical consequences from first principles and to design a well-ordered, logical system.

Duhem notes, but without really developing the point, that the mechanical models of the British physicists were part of a larger proclivity for representation via detail and individual units. His analysis articulates the well-worn conflict between Anglo-Saxon empiricism and French rationalism that dates back to the two founding philosophers of modern science, Francis Bacon and René Descartes. In early seventeenth-century England Bacon published several treatises that provided a philosophical basis for and defense of the new experimental and mechanical methods through an inductive approach based on experiment and observation. In the first book of the *Novum Organum* he advocated a method that "constructs its axioms from the senses and particulars, by ascending continually and gradually, till it finally arrives at the most general axioms, which is the true but unattempted way" (1952:108; I.19). In contrast, Descartes's method employed deduction from first principles and then tested deductions with experiments. Descartes demonstrated the power of his approach with contributions to the fields of optics and meteorology. He also developed a new synthesis of the fields of geometry and algebra that became known as analytic geometry and that laid the groundwork for much of modern mathematics.

At one level the cultures from which Descartes and Bacon emerged had a great many similarities. In both kings were strong and religious orthodoxy was sanctioned through persecution of heterodoxy. Likewise, both Descartes and Bacon were, in a sense, apologists for a religious order even though they were reforming aspects of it. However, there were also crucial differences between the two lives and cultures. Bacon advanced rapidly under James I and became lord chancellor of England; he was therefore part of high-level political circles, that is, until he fell from favor for accepting bribes. In short, for most of his adult life Bacon was part of the official order in a time prior to the English Revolution. Nevertheless, he was still writing in a culture that over a period of several centuries had been developing a parliamentary tradition and a common-law legal system. Furthermore, by the early seventeenth century Britain had become a largely Protestant country in which numerous, more democratic versions of Protestantism were emerging in opposition to the structures and practices that Anglicanism retained from Catholicism. Thus, in politics, law, and religion, British society was developing a "bottom-up" culture that loosely resonated with Bacon's empiricism. Although Bacon worked closely with King James I and was no friend of parliament, he was immersed in a culture that was developing what could be called an inductive style in a variety of cultural domains.

In contrast, France remained much more of a hierarchical society, with political authority defined around a highly centralized state and official religious authority centered on the Pope. Like Bacon, Descartes was somewhat at odds with his culture: he rejected his Jesuit education and found the Protestant country of the Netherlands more congenial. Nevertheless, just as English culture might be described as having an inductive style, so French culture might be described as having a deductive or "Top-down" style. As authority in French society ran from the king or pope downward, so Descartes constituted a system that ran from first principles to empirical observations.

More generally, Anglo-Saxon culture tends toward individualistic representation and inductive logic. Anglo-Saxon psychology was individualistic, and its social theory was economistic, both of which rested on details. Likewise, as Duhem noted, Anglo-Saxon novels and theater tended to go into details and individual character traits to such an extent that his French mind found the narratives boring. Thus, the patterns of a scientific style are part of a general cultural style: a physics of mechanical units, a psychology of associated ideas, a social theory of individual rights, an economics of utilitarian calculus, and so on. All were part of the same cultural logic that produced a literature of details and individual characters, a legal system based on common law, a religious system based on sects, and a state based on guarantees of individual freedom.

French physics, literature, social theory, law, religion, and government were in many cases more hierarchical and deductive. The empiricism/rationalism dimension is really a special case of a value difference between two cultures. It is true that both France and England were modernizing in different domains at different speeds. Furthermore, French culture of the twentieth century is in many ways more democratic and modern than seventeenth-century English culture. Nevertheless, at any given time there is a general tendency for the French to think and organize their social and intellectual life along deductive lines, at least relative to the British tendency to do so along relatively inductive lines. Many great pairs of scientists and thinkers across the continental divide—Darwin/Lamark, Dalton/Gay-Lussac, Rumford/Carnot, Galton/-Quételet, Ampère/Faraday, and so on—have yet to be discussed comparatively from a cultural perspective.

In social theory the differences are perhaps most clearly articulated. The English philosophers Locke and Hobbes, for example, founded their theories of society on a social contract, whereas the French philosopher Rousseau argued that the social contract itself was meaningless unless it was preceded by a sense of a community or common self. Here, Locke deduced society from a meeting of individuals, whereas Rousseau pointed to a social level that had to precede the social contract. That basic difference was perpetuated into the nineteenth century in various French and British formulations of society. For example, John Stuart Mill placed psychology as the queen of the sciences in a society run according to individual rights, whereas his contemporary Auguste Comte placed sociology at the apex of a society run by an elite. Thus, Mill looked at society more from the individual's perspective, whereas Comte (and especially his intellectual heir Durkheim) tended to look at the individual from society's perspective. The debate continues in the twentieth century in the form of resistance to and reworking of Durkheim's sociology and Lévi-Strauss's structuralist anthropology. One might label the difference in intellectual styles between the Anglo-American and the French empirical/rationalist, inductive/deductive, facts/truth, or individualistic/holistic. Whatever the label, each generation seems to recreate—but also rework—the divisions in a different intellectual or scientific space.

As scientific communities and national societies have become more internationalized in the twentieth century, the analysis of national style differences in the construction of scientific theories may provide fewer insights. Even at the turn of the century Duhem noticed that the "English model" of physics was diffusing to the continent. Nevertheless, in the social sciences and humanities there are still distinctively continental and Anglo-Saxon styles, and occasionally some debates in the natural sciences—such as French versus American claims over the AIDS virus—suggest that national cultural differences may still play an important role in the natural sciences as well. A failure to investigate more carefully such differences in the national sciences and social theories amounts to buying into the ideology of science as a supranational phenomenon that is everywhere the same.

The Production of Technototems: From Bricolage to Reconstruction

So far my discussion of technototemism has been limited to showing how technical distinctions are coproduced with social distinctions. The process is further complicated when new social groups receive and rework technical knowledge. The anthropologist Claude Lévi-Strauss introduced the term *bricolage* to describe a similar process by which the first natural theories, myths about nature, passed across social groups. The *bricoleur* is a jack-of-all-trades who takes whatever is at hand—pieces of wood, metal, spare parts, junk—and reassembles them to build new objects or to fix old ones. The French anthropologist argued that people in one culture take the myths of other cultures and retell them by borrowing the elements of those myths—the characters, settings, events, and so on—and combining them in a different order. The elements remain largely the same (although there are sometimes substitutions), but they are reorganized in different ways.

Does bricolage apply to science and technology? If one wishes to maintain that science is transcultural and everywhere the same, then the answer would have to be no. That is the assumption in *The Savage Mind*, where Lévi-Strauss contrasted the bricoleur and the engineer. Because bricoleurs take what is at hand and reassemble it, he argued, they tend to work from whatever patterns flow from their materials. In contrast, engineers work from first principles. Here, Lévi-Strauss associates the deductive method with science, engineering, and the "modern," whereas the inductive method is linked to jacks-of-all-trades and the myth-makers of non-Western societies.

The distinction between the engineer and the bricoleur can be challenged on both theoretical and empirical grounds. One argument is that there is a cultural bias in Lévi-Strauss's construction of the engineer and the bricoleur. Lévi-Strauss's engineer operates deductively from first principles in a manner that follows the French style of doing philosophy and science. In contrast, Lévi-Strauss associates the empirical and inductive style with the "primitive." The association is ethnocentric on a number of levels. Not only does it fail to grant to indigenous peoples the ability to work out deductive systems of thought, but it also associates the British (and Anglophone) style of doing philosophy and science with the bricoleur/primitive. Furthermore, although the engineer and bricoleur are both masculine figures, the ways in which rationality is assigned to one side of the split suggest a parallel with Western patterns linking rationality and gender.

Lévi-Strauss fell victim to a more general error that involves the presumption of a single transcultural scientific and technical knowledge. More widely accepted today is the idea that science and technology also operate according to the patterns of the bricoleur. For example, the philosopher Jacques Derrida argues, "If one calls *bricolage* the necessity of borrowing one's concepts from the text of a heritage which is more or less coherent or ruined, it must be said that every discourse is *bricoleur*" (1978:285). Likewise, empirical studies of laboratories show that scientists are much more like the tinkering bricoleurs than Lévi-Strauss had imagined.

Although Lévi-Strauss's construction of bricoleurs is flawed, I have returned to the idea because it helps move the discussion beyond the analysis of technototemism, that is, the cultural interpretation of the ways in which different experts, communities, and so on construct different versions of science and technology. In constructing different versions, scientists and engineers are sometimes engaged in a process akin to bricolage. In other words, they are taking the versions of other communities and *reconstructing* them so that the elements are recombined to better fit with their own local culture, which includes their perceptions of what construction best matches their interests. I therefore enlist the concept of bricolage for a specific purpose. To return to the culture-politics theme, it is not enough to provide a purely structural or interpretive analysis that carefully follows out the threads of the cultural meanings of technosocial difference. Becoming aware of the history of scientific theories and technological practices as totems for race, class, gender, and other social divisions is only the first step. The next step is to look for ways in which different groups actively reconstruct sciences and technologies by positing alternatives consciously linked to their social identity. In some cases these alternatives open up possibilities for finding new ways of doing science and technology that can promote greater social justice. As a result I prefer the term *reconstruction* to bricolage in order to signal the level of actors, strategies, and power relationships.

Before examining some cases of reconstruction, it is important to point out that I am not arguing that women and members of underrepresented social groups are always engaged in reconstruction. In fact, most members of historically excluded groups who are in the technical professions are constructing knowledge and technology in the dominant traditions in which they were trained. They are sometimes rewarded for their efforts and can serve as role models, as in the case of the women portrayed in *Nobel Prize Women in Science,* by the STS researcher Sharon McGrayne. Likewise, men from dominant groups can also reconstruct technoscience from other perspectives. In other words, a position as a member of the dominant groups does not always correspond to a stand in favor of those groups, just as lack of membership does not always correspond to a stand counter to the perspective of those groups. What I am interested in, then, is a narrow subset of cases in which women and people from nondominant groups have actively reconstructed science and technology from a perspective that is explicitly linked to their outsider identity. Although all reconstruction is construction (and vice versa), I prefer to use the term *reconstruction* to refer to a particular type of construction by groups and individuals who have perspectives different from the mainstream of technical expertise.

Women's Reconstructions

There is now a growing literature that provides examples of ways in which women scientists and engineers not only have unpacked the ideological nature of science and technology but have also built alternative technototems of their own that explicitly inscribe alternative social visions. I will consider two examples here, one from primatology and one

from automotive engineering. The popular image of women and primatology is shaped by movies and documentaries on Jane Goodall and Dian
Fossey. They became known for living among chimpanzees and gorillas
and for bringing about new scientific insights based on their long-term,
close-contact studies. However, as women entered primate studies and
physical anthropology, they changed it in a number of ways that are not
well represented in the popular images. Many of those changes are analyzed by the historian Donna Haraway in her books *Primate Visions* and
Simians, Cyborgs, and Women, two of the most influential volumes in the
making of a feminist-cultural studies-anthropological version of STS in
the 1990s. I will focus here on one of the changes made by women primatologists and physical anthropologists in the late twentieth century:
the introduction of a theory that has sometimes been called "woman-the-
gatherer." In addition to serving as an example of reconstruction, the
case also helps to resist popular misrepresentations associated with
women primatologists as icons of maternal methods.

The woman-the-gatherer theory posited an explanation of the origins
of humanlike (hominid) species from apelike (hominoid) ancestors, a
transition that primatologists and physical anthropologists have generally assumed to be linked to a change in habitat from forests to savannahs. By the 1960s the dominant explanation of that transition was the
so-called man-the-hunter theory. As a first approximation, the complex
theory can be described as linking hominid bipedalism to evolutionary
selection pressures that favored species variations that could cover
greater distances for hunting in the savannah. The theory also assumed
a sharp sexual division of labor in which males did the hunting and
females stayed at "home" to raise the children, not unlike the sexual division of labor of many Western societies during the time the theory was
produced. Although in retrospect the theory may seem suspiciously and
transparently biased by a masculine perspective, it is important not to
paint its principle architect, the anthropologist Sherwood Washburn, as
a "neanderthal." Haraway notes that Washburn's generation of physical
anthropologists actively developed theories that contested the older
generation of evolutionary theories, which had constructed racial hierarchies (1989:87). Furthermore, Washburn is known for having trained a
disproportionately large number of women physical anthropologists in
comparison with male colleagues of his generation.

One of Washburn's students is the anthropologist Adrienne Zihlman,
who, at the annual meeting of the American Anthropological Associa-

tion in 1970, heard a paper entitled "Woman the Gatherer: Male Bias in Anthropology" by the anthropologist Sally Linton Slocum (1975). Zihlman and the anthropologist Nancy Turner developed the woman-the-gatherer theory during subsequent years. Although there are various versions of the theory, it is sufficient for the purposes here to work with a first approximation. Research by Goodall and other primatologists had revealed how forest-dwelling chimpanzees construct primitive tools and also hunt and kill small animals. That research made more problematic the previous links among tool use, hunting, and the transition to the savannah. Furthermore, a number of studies pointed to the complexities of primate social structure, in which complex mother-infant bonds, female sexual choice, and matrifocal social structures were prominent. Studies of the !Kung hunter-gatherers in the African savannah, which were believed to be close to early hominid societies, also showed that the bulk of food calories came from gathering rather than hunting. Thus, a number of research lines converged to suggest that the story was more complex than man-the-hunter. As Haraway summarizes:

> Zihlman argued that what was new was, precisely, extended generalist gathering by both sexes, in which the productive, exchange, and reproductive activities of the females would dynamically propel, but not uniquely cause, a species-making transition. Extended duration of mother-young relations, improved female sexual self-determination, greater complexity of matrifocal groups requiring greater social skill in both cooperation and competition from both sexes, and cultural innovations (tools, social arrangements, and other technologies) enabling these extensions were seen as originary. (1989:339–40)

One might expect that, given the mounting evidence, Zihlman's more inclusive woman-the-gatherer theory would have easily displaced the older model. However, even though Zihlman resisted formulating a mirror-inversion of man-the-hunter, the woman-the-gather theory was often interpreted that way and then rejected as having a reverse gender bias. (When I explain the theories in my classwon, some of my male students usually have their hands up before I am even done to point out a perceived "reverse" gender bias.) Others accepted gathering as important, but assigned the innovation to men! Yet others, including Washburn, minimized the contributions of Slocum, Tanner, Zihlman, and colleagues by claiming that gathering was already included as a contributing factor in the man-the-hunter theory. In other words, woman-the-gatherer became reincorporated into a somewhat expanded man-the-hunter theory.

As gendered theories (or technototems), woman-the-gatherer and man-the-hunter provide one example of the cultural politics of what happens when a previously excluded group enters a field of inquiry. Many will continue to work on projects that are not controversial to the leaders in the field, but in some cases some members of the new groups may bring about major changes in theory or methodology. However, the dubious fate of woman-the-gatherer shows how reconstructions of science by previously excluded groups can be reincorporated or reappropriated—that is, re-reconstructed. Thus, reconstruction is an ongoing process of negotiation and renegotiation. To underscore the point, it is perhaps useful to consider another case, one that involves the reconstruction of technological design rather than scientific theory.

When the Ford design executive Mimi Vandermolen was given the power to redesign a car, she consulted fifty women about problems they faced in automobiles. The newly designed Ford Probe resulted in a number of changes that made the car more appealing to the professional women whom Ford targeted as their market. For example, gas pedals were redesigned to be at a right angle for women who use high heels, knobs were contoured so that even women with long fingernails could clear them easily, and a new molding was used for the seats, so that women in dresses could get in and out of the car more easily. Furthermore, the cowl (the point at which the windshield meets the engine compartment) was lowered because their tests showed that women generally like to sit high and close to the wheel, and they preferred the lower design. As a woman colleague and Probe owner also pointed out to me, the car is available with a panic button on the key chain. If someone appears to be stalking the car owner, the owner may hit the panic button, and immediately the lights of the car will start flashing and the horn will start honking.

The redesigned Ford Probe was based on the results of interviews with women at Ford and on discussions in a marketing committee within the company. As a result, the issues that were salient to these women reflect the clothing and comfort concerns of women who work for a large American corporation, and mostly professional women within that category. The design changes can hardly be considered radical, and they inscribe in the technology a particular image of women that could be considered patriarchical. In other words, the car is designed for a woman who has long fingernails, uses high heals, and so on. Thus, although the Probe provides one example of how a technology can be redesigned in light of masculinist design biases, the new design may also inscribe new biases of its own.

It may be worth reiterating the point that the reconstruction of science and technology involves more than just eliminating biases. The concept of eliminating biases tends to go hand in hand with the assumption that eventually it would be possible to achieve an ideal state in which science and technology are divorced from the sociocultural. The alternative model that I and many STS researchers use assumes that science and technology will always take the form of totemic relationships. The question becomes one of figuring out what biases are built into the technototems, and then building new theories and devices that provide more equitable reproductions of society. However, the end result is not a theory or technology that is outside culture. Instead, it is merely infused with different cultural politics, some of which we may not even be aware.

An African American Science

An example of an alternative science constructed from an African American standpoint is the work of George Washington Carver. A largely rejected figure in African American studies, Carver has frequently been criticized for not challenging traditional white stereotypes of the ideal African American male as a humble, servile person. For example, one African American scholar argues that a better model for an African American scientist is the aforementioned Ernest Everett Just, who openly supported racial equality and did not play humble pie to the whites (Winston 1971:702–4). Furthermore, whereas Carver did not publish research papers in refereed scientific journals, Just was well published, and yet he still suffered great prejudice. However, Carver remains an interesting example because the content of his scientific research was deeply infused with his concern for poor farmers in the South, particularly African Americans, who still labored under the legacy of King Cotton and slavery. Thus, unlike Just, Carver provides a model not of an African American scientist but of a producer of an African American science. In many ways his project was a failure (just as the Ford Probe involves traditional notions of womanhood even as it builds a woman's car), but the Carver case is instructive because it focuses on the question of reconstructing content.

The early biography of Carver is fuzzy, but he is said to have been born a slave. His owner, Moses Carver, was pro-North, and for his allegiances he suffered attacks from southerners during the Civil War period. At one point George and his mother Mary were kidnapped and literally "sold downriver." Moses Carver never found Mary, but in an exchange he eventually reclaimed George, who later became known as the "man who was traded for a horse." The kidnapping incident is only one example of the difficult childhood and youth of George Washington Carver, whose formative years were filled with events demonstrating the oppressive hand of racial prejudice and an equal determination to overcome obstacles.

Eventually, Carver encountered some support from a sympathetic white couple who recognized his talents, and he was able to complete his education and enter Iowa State College of Agriculture and Mechanical Arts in Ames, Iowa. At that time the school was a place of great intellectual ferment as a center for emerging agricultural science in the United States. The faculty included Dr. Louis H. Pammell, perhaps the country's leading botanist; James Wilson, the Dean of Agriculture and director of the experimental station who became secretary of agriculture under Presidents McKinley, Roosevelt, and Taft; Wilson's assistant, Henry C. Wallace, who became the secretary of agriculture under Presidents Coolidge and Harding; and his son Henry Wallace, who grew up taking nature hikes with Carver and later became the secretary of agriculture under Franklin D. Roosevelt and his vice-president in 1940. Iowa therefore provided rich soil for a great mind, and the progressive-minded people there recognized his talents. After Carver received his B.S., the faculty invited him to stay as assistant botanist at the experimental station and director of the greenhouse.

In 1896 Carver's life changed dramatically when he received a letter from Booker T. Washington, who invited Carver to join him at the Tuskegee Institute. Washington, like Carver, was to remain a controversial figure among African Americans, for both adopted a strategy of appeasement that irked the more militant and political African Americans in the urban North. (In a way the criticisms are an early manifestation of what is today sometimes called the "Martin/Malcolm" split—

the differences between Martin Luther King, Jr., and Malcolm X that symbolize the tensions in African American politics between its more southern, rural, conservative forms and its more northern, urban, radical forms.) Although the northern criticisms of Washington are perceptive, he was also a very shrewd leader who knew how to play politics among southern conservatives and northern liberals in order to secure funding and support for his bold attempt to build higher education for African Americans in the rural South. Furthermore, under Washington's protection Carver was able to pursue teaching and research, which allowed him to dedicate his life to helping the poor, rural, African American farmer.

Among Carver's achievements he provided models of science, research, and education that are still applicable today in rural areas throughout the world. For example, Carver founded the "moveable school," a wagon that traveled around and met with poor people, usually outside churches on Sundays, to give them suggestions on improving agriculture, carpentry, sanitation, and so on. Perhaps most well known of his achievements was his work on applications for alternatives to King Cotton. In the first decades of the twentieth century the boll weevil was moving in from Texas and destroying cotton crops. Carver took advantage of the blight to encourage poor farmers to plant peanuts, a nutritious and hardy crop. Many farmers did indeed follow his advice, but their decision soon resulted in a glut of peanuts. Although planting peanuts had the advantage of getting the small farmers out from under the thumb of King Cotton, they still needed a cash crop.

Carver answered the problem by developing nearly three hundred products for the peanut and over a hundred for the sweet potato. Many were food products such as peanut milk, which missionaries subsequently took to Africa and used to save children in places where cattle could not survive. Carver also developed industrial and dehydrated products that were decades ahead of their time, and his inventions helped provide the basis of the southern peanut industry. He also testified before Congress to help protect the nascent peanut industry from a foreign invasion of lower-cost Asian peanuts then flooding the market. His fame reached national and even international levels, and he received many offers for high-paying research positions. Thomas Edison, for example, offered Carver one hundred thousand dollars to come work in his Menlo Park Laboratory, and Henry Ford tried to get him to come to Dearborn, even going so far as to set up a laboratory for him.

As a genius figure, Carver became a legend in his own time. He retained a saintly persona, wearing the same coat for almost thirty years and leaving paychecks uncashed in his room or giving them away to students in need. He epitomized an image of the humble Negro that racist whites found appealing. He gave many speeches before southern industrialists, even when he was sometimes not allowed to eat with them or had to enter through the servants' quarters. He saw himself as sacrificing his own pride for the goal of helping poverty-stricken farmers. In discussions of his research methods he downplayed his work as "Talking to plants" and his own scientific acumen as merely God's work. (However, as students were to discover, he was actually a very meticulous methodologist, which underscores the point I made above about the dangers of essentializing women and, in this case, African Americans as always being associated with different methods.)

Although successful as a model for African Americans who had to battle racist beliefs about their lack of intelligence or inability to do science, as a political figure Carver was a failure. He failed to speak out on civil rights issues and become the Frederick Douglass of African American science. However, he did succeed in helping to open the door for an alternative to King Cotton. In this respect he was not just another brilliant scientist who happened to be African American; Carver infused his science with his social concerns and built an African American science. He posed one totem against another—the peanut (and sweet potato) against cotton—and he built up an alternative science and technology that could benefit those who farmed the alternative plants. He also manipulated the system to the best of his abilities for his own goals of reconstructing a new agricultural science and new southern society. Of course, his sociology and politics were naive, and large, white capital appropriated his research while white politicians appropriated his fame. To paraphrase Karl Marx, men and women make their own history, but not exactly as they please. Nevertheless, Carver provides an example—even if a partial failure—of the levels to which one can take the idea of the reconstruction of content.

National Reconstructions

Scientists in different countries also engage in the reconstruction of scientific traditions as they receive them from other countries. For example, in *Psychoanalytic Politics* the STS researcher Sherry Turkle compared the reception and reworking of psychoanalysis in France and in the United States. In the United States social conditions set the stage for the early acceptance of psychoanalysis, but the reception that Freud described as overly enthusiastic carried with it a radical reworking that reoriented psychoanalysis to a message of hope in a culture of pragmatic self-improvement. Americans rejected much of the critical and historical side of psychoanalysis, and instead they reconstructed psychoanalysis in a medicalized mode that was oriented toward the pragmatic issue of cure.

In contrast, in France there was an existing research tradition of dynamic psychology associated with the work of Pierre Janet, so Freud, who early in his career had studied in France, was seen as less novel or even derivative. The psychiatric and medical establishment there rejected psychoanalysis, which remained largely restricted to intellectuals, artists, and writers. It was only after the political events of May 1968 that psychoanalysis began to achieve the prominence in French culture that it had long occupied in American culture. Just as Americans had transformed psychoanalysis, so the French rewrote the Austrian science. Under the leadership of Jacques Lacan, "French Freud" became a new science that was in many ways the opposite of psychoanalysis in the United States. In Turkle's words,

> The Lacanian paradigm is structuralist, emphasizing the individual's constraints rather than his freedoms; it is poetic, linguistic, and theoretical rather than pragmatic, and tends to open out to a political discourse which raises questions beyond the psychoanalytic. (50)

In short, whereas in the United States psychoanalysis became part of the medical establishment and Freud for a long time was interwoven in sexist and heterosexist politics, in France psychoanalysis became associated with the left and with French feminism.

One might reply to this argument as follows: Well, even if human sciences are reconstructed as they pass across cultures, certainly the same cannot be said for the natural sciences. In fact, there is growing evidence for the existence of reconstructions in the natural sciences as well. For example, in *Beamtimes and Lifetimes* the anthropologist Sharon Traweek has shown that the Japanese funding system provides physicists with large initial outlays but little subsequent funding (1988:46–83). Furthermore, the system of examinations in Japan has prevented the formation of a cadre of experienced, highly trained technicians. As a result, when the Japanese physicists developed their own machines for measuring high-energy particles, they opted for reliable, durable detectors that were built with the best available technology. In contrast, at least one American detector—that of the Stanford Positron Electron Asymmetric Rings—is in many ways the opposite of the Japanese style; the American detector was built with an open architecture that allowed the physicists to rebuild the machine. Whereas the Japanese machine was designed for precision at the expense of new data, the American detector was designed to make it possible to search for a wide range of unexpected particles.

One might still argue that the content of the physicists' theories, rather than the style of the detectors, is more or less the same in Japan and the United States. At one level the argument is almost tautological, for by definition science is a social machine for producing knowledge that can be accepted across diverse cultures. As scientific communities move from local research schools and national enclaves to international networks with international meetings, ideas and theories flow quickly from place to place and scientists live increasingly in transnational disciplinary cultures. Nevertheless, some studies now show how the content of theories is at times reconstructed in different national cultures, even in the natural sciences. For example, in the essay "The Bio-Politics of a Multicultural Field," Haraway discusses some ways in which primatology has been reworked in Asian settings (1989:244–75). Westerners tend to operate in terms of a sharp division between nature and culture, and they locate primates in a nature that is seen as wild. In contrast, the Japanese tend to view nature as something to be cultivated by the human hand (as in Japanese gardens), and they tend to view the animal/human relationship more in terms of a family metaphor in which animals are the humans' younger siblings. Those different assumptions played themselves out in different formulations of primatological methods. For example, Japanese primatologists tended to feed their primates, a procedure that made it easier to observe the animals. In contrast, Westerners often rejected provisioning in favor of methods that would have less chance of disrupting the behavior of the animals in their natural state. Furthermore, the Japanese also tended to emphasize an empathetic method that stressed the more human qualities of the monkeys as individuals and members of social groups.

At the theoretical level Japanese primatologists also developed alternatives to Western primatology by reporting that Japanese macaques were organized as a matrilineal society and by emphasizing—long before feminist revisions occurred in the West—the importance of males' relationships to female hierarchies. Although Japanese society is highly patriarchal, men are also more willing to recognize the importance of mothering and mother-son relations in daily life, as well as female deities and shamans in the Shinto religion. Once again the cultural context contains the conditions of possibility for developing alternative theories.

Haraway also discusses how India represents yet another possibility. In India monkeys belong to a sacred supernature rather than to a cultivated nature, as in Japan, or to a wild nature, as in the West. One of the gods in Hinduism is the monkey Hanuman, who is portrayed rescuing a human deity. In both the religious story and in everyday life monkeys are portrayed as interacting with humans. The cultural meaning and social relations with monkeys in India form the backdrop to the Indian government's ban on exports of monkeys to the U.S. The ban occurred when Indians discovered that the American government was subjecting thousands of monkeys to radiation experiments to better understand how people would function (or malfunction) in the event of nuclear war. In addition, Indian primatologists have performed studies on the interaction of monkeys and humans, including a survey that documented the decline of monkey populations. Whereas Westerners have focused on studying primates in the original state of nature, Indian primatologists have drawn on their cultural background, which constructed monkeys more in relationship to humans, to develop a different emphasis in primate research.

These examples show how easily relationships of national difference and national style slide into issues that are colored by neocolonial and international politics. Indian and American theories are different, but the differences can be expressed in a play of power by which one side may try to stop the other side from pursuing science from within its framework. Wealthier countries may fund third world research that fits their ideas and third world countries may pull up the welcome mat for researchers from wealthier countries. In this context it is easy for intellectual and theoretical differences to slide quickly into accusations of imperialism or national chauvinism.

Conclusion: Technototemism, Reconstruction, and Power

I have used the metaphor of totemism for a type of analysis that reveals cultural and political difference where previously only technical difference was evident. That analysis does not seek to explain technical difference by social difference; instead, it shows how the two are interwoven or conconstituted. Distinctions in one field, the technical, are made meaningful because they are translatable into distinctions in another field, the social, and vice versa. Furthermore, I have argued in favor of a second step that pinpoints the processes by which the content of scientific knowledge and technological designs is reconstructed from the standpoint of groups that historically have not had the same access to the means of technoscientific production as have upper- and middle-class, educated men of European descent. In the process the analysis can point to alternatives and choices where previously only one possibility may have been envisioned. Cultural-political analysis also points to what is at stake: who is connected with what side of a technical choice or a debate. Yet, in a sense the story only begins with these clarifications.

To map out fields of social and technical difference is only a first step toward analyzing how those differences are embedded in power struggles over whose theory or device is considered more true or more successful. In STS one influential approach for analyzing these power struggles is the work of Bruno Latour, Michel Callon, and their colleagues. Their approach, sometimes referred to as actor-network theory, views the outcome of scientific and technical conflicts as largely determined by an ability to get others to go along with a version of what counts as the "Truthful" or "successful." Scientists and engineers—the actors—build heterogeneous networks of colleagues, backers, machines, publications, organizations, and so on to make their theories and devices successful and unassailable. As those networks expand, facts become more factual and technologies become more successful. In other words, truth and success are socially negotiated. Scientists, like entrepreneurs, need to convince other people to back up their work.

The approach of examining science and technology as an active process of construction is a very powerful way of demystifying questions of how widely accepted knowledges and technologies came to be accepted. However, the case studies of technototemism—particularly those associated with class, gender, sexuality, race, ethnicity, and nationality—suggest that by itself the image of growing and competing actor-networks is far from complete. Actors come to their networks with histories, and those histories are embedded in longstanding social structures that include patriarchy, racism, colonialism, and class domination. For example, as the historian Donna Haraway shows, those primatologists who wished to battle the advocates of the man-the-hunter theory were generally women who were attempting to gain a place for themselves and their work in patriarchal academy and society. It is possible to describe their theories as flowing through alternative networks that they built, but a theory of technoscientific reconstruction would still need to address the crucial point of how the playing field for actor-networks is far from level. Those who control the networks of science and technology are not likely to let in theories and devices invented by those who do not. If they do, they are likely to do so with only minor modifications. For example, the Ford Probe hardly put any of the deeper structures of patriarchy or other forms of domination into question. Those who advocate more serious alternatives—public transportation or (in the case of King Cotton) land reform—most often suffer the mechanisms of marginalization.

Actors come to networks within cultures that provide them with biases about appropriate forms of knowledge, methodology, and machinery. Thus, although an actor-network analysis brings to STS the helpful corrective that shows one way in which structural change is possible, discussions of actor-networks need to be framed by an analysis of culture and power. The frame can show how successful technoscientific reformers—those who build new cars or remake the peanut—can also end up reproducing the structures they challenge. The frame can also bring out the processes of what the STS researcher Brian Martin has called "intellectual suppression" (1986). In turn, after recognizing that the rules are the game are rigged, one can begin to explore what I have called "strategies of circumvention," that is, successful ways of reconstructing science and technology that are also cognizant of the structures of culture and power in which one is embedded.

3 | The Origins of Western Science: Technototems in the Scientific Revolution

Most stories about cosmopolitan science and technology begin with the scientific revolution, that is, the events in Western Europe between 1500 and 1700. Much of the science and technology practiced throughout the world today draws on the basic principles, formulas, and concepts that were elaborated, if not developed, in Europe during that period. In this chapter I examine the cultural politics of narratives of the "scientific revolution" as a myth or origin story. In the process I provide one example of how to do reconstruction, in this case, the reconstruction of the history of the scientific revolution.

What Was the Scientific Revolution?

The Legacy of Ancient Science

Stories of the scientific revolution usually begin with the legacy of the ancient world. Ancient science is generally associated with the Greeks, although increasingly the Greeks are seen as the inheritors of scientific traditions from Babylonia, Egypt, and other cultures. Many of the ideas about the natural world that comprised the accepted worldview in Europe in 1500 can be traced to natural philosophers of the ancient world such as Ptolemy, Aristotle, and Galen. However, as historians now

FIGURE 3.1
Retrograde Motion and Geocentric Theory

Mercury   Mars

Seen against the backdrop of the stars, and over a period of days or weeks, the planets often moved in odd trajectories.

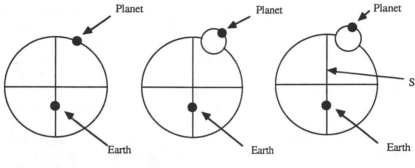

Eccentric. The planet's orbit is a circle around an "eccentric" center, which is slightly off from Earth.	Eccentric with epicycle. The planet also moves in a small circle as it moves around the eccentric center.	System with equant. The planet now moves at a uniform rate around point S, rather than the eccentric center or Earth.

recognize, older stories of the "Dark Ages" in between tend to filter out Arab and non-Western influences on the inherited medieval worldview, just as they downplay the ferment and developments of medieval science in the centuries preceding 1500. Nevertheless, I shall begin with the ancient Greeks because their science is the usual starting point for discussions of the scientific revolution. Their "science" soon reveals itself to be very different from that of the modern period. As historians often say, the past is another culture.

In the ancient worldview the earth was at the center of the universe, and the sun, moon, planets, and stars all revolved around the earth. The view is an understandable conclusion that can be drawn from even a few quick looks at the sky, but it also rested on some very careful observations of the planets. For example, against the backdrop of the stars the path of Mercury or Mars engages in what is today called retrograde motion (see figure 3.1). At a certain point the planet appears to stop, go backward, and then zigzag forward. The Ptolomaic or geocentric theory assumed perfect circular motion of the planets and the sun around the earth. As a result, retrograde motion was explained by saying that a planet's orbit around the Earth is in the shape of a circle that contains a little circle known as an epicycle. One can achieve almost any kind of retrograde motion by combining epicycles with eccentrics and equants. Eccentrics are orbits around a point that is not centered on the earth, and equants are points around which the planet moves in uniform speed (see figure 3.1 again).

In today's sun-centered system, scientists explain retrograde motion as an optical illusion created by the relative movement of the orbits of the Earth and planets around the sun. As the planets move in their elliptical orbits, the relative movement of the Earth and either an inner planet such as Mercury or an outer planet such as Mars creates the effects of retrograde motion (see figures 3.2 and 3.3). The ancient Greeks and medieval philosophers were aware of alternative, sun-centered theories of what we call today the solar system. However, as long as they assumed that all orbits were circular (rather than elliptical, as Tycho Brahe later claimed), Ptolemy's system was the only one elaborated mathematically that fit the observations. His theory also fitted with common sense and the idea that heavenly bodies all took the shape of perfect circles.

A second argument in favor of the geocentric model had to do with motion and the behavior of objects on the surface of the earth. The ancient Greeks argued that if the earth rotated (which it would have to do in a heliocentric system), then there would be a tremendous wind and the birds would get left behind. Likewise, the water of the oceans would probably slurp over the shorelines. It made much more sense to assume that the world stayed put.

A whole science of natural motion was constructed around the idea that the earth was at the center of the universe. Aristotle's theory of natural motion was based on the assumption that all material in the universe was composed of four basic elements: earth, air, fire, and water. Materials composed of earth and water had a natural propensity to return to the earth and, therefore, to fall, whereas those composed of fire

FIGURE 3.2
Retrograde Motion for an Inner Planet

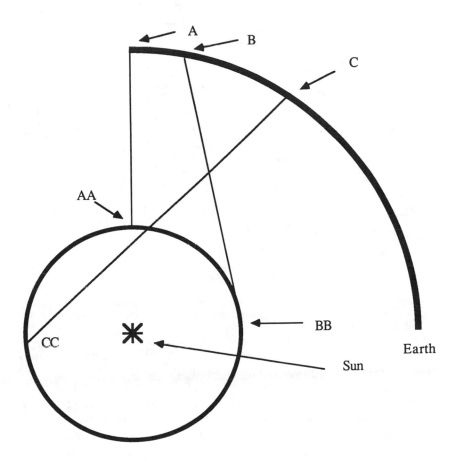

When Earth is at point A and the inner planet (e.g., Mercury) is at point AA, the inner planet appears to be going in one direction. The inner planet then appears to reverse direction
when Earth arrives at point B and the inner planet is at point BB. (Not to scale.)

FIGURE 3.3
Retrograde Motion for an Outer planet

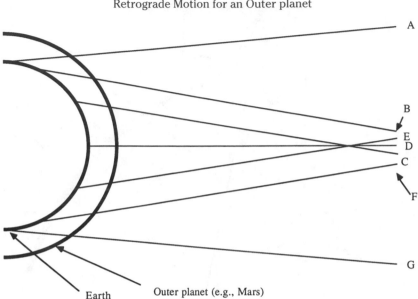

Against the backdrop of the stars, the outer planet appears to go forward until reaching point C, then it moves backward to point E, then proceeds forward again through points F and G. (Not to scale.)

and air had the opposite tendency, to rise. Earthbound materials only moved if something propelled them, such as a horse pulling a cart. Once the horse or other force stopped propelling the object, it would come to rest or fall immediately to the ground. For apparent anomalies such as an arrow shot by a bow, Aristotle argued that the arrow pushed the air in front of it, which then rushed around the arrow to fill in the vacuum left behind it, thus propelling the arrow forward. Likewise, to explain why an object accelerates as it falls, Aristotle argued that objects rushed more quickly as they got closer to their natural resting place. He gave the rather implausible explanation that involved attributing to objects a desire to return to their natural resting place. His followers explained acceleration by arguing that there was a shorter column of air beneath the object, and as a result there was increasingly smaller air resistance. Thus, the theories of ancient physics were plausible and coherent, even if scientists today would consider them mistaken.

Ancient natural knowledge also had well-developed systems of medicine, physiology, and anatomy. For example, Galen practiced dissection and had a sophisticated sense of human anatomy. Questions of physiology such as the circulation of blood, however, remained more open to debate. Modern science has shown that blood circulates from the heart through the arteries and back to the heart through the veins. In contrast, Galen argued that there were two kinds of blood: one ran from the liver through the veins and provided nutrients to the organism; the other ran from the heart and contained a life-giving material called pneuma. For Galen circulation occurred independently in the two systems through an ebb and flow within each. The only interaction between the two systems was that the blood from the veins seeped through the dividing wall of the heart into the left ventricle, where it was purified to become part of the arterial blood. However, the heart did not pump blood; rather, it drew blood inward, and it also drew in air directly from the lungs.

The Story of the Scientific Revolution

At the beginning of the sixteenth century ancient teachings such as those of Ptolemy on the planets, Aristotle on motion, and Galen on physiology constituted the basis of accepted wisdom. In schoolbook stories of the scientific revolution, a few bold and courageous thinkers in sixteenth- and seventeenth-century Europe rebuilt the entire edifice of ancient theories on nature and the universe. However, for some time now historians have questioned the extent to which there was really a "scientific revolution" rather than "evolution" of the ideas of previous thinkers. For example, in the centuries prior to the scientific "revolution," a number of scholars, particularly in Oxford and Paris, questioned central aspects of Aristotelian physics and began to develop a theory of what we would today call inertia.

The history of the scientific revolution is therefore a construction, as are all historical narratives. Yet, this particular history may have more fictive qualities to it than others. To begin, the title of the story is fictive. The term *science* is an anachronism, because the word *scientist* did not emerge in today's sense until the nineteenth century. The term used at the time was *natural philosophy*; however, the phrase "natural philosophy revolution" just does not seem to have the same ring to it. As for the word *revolution,* the plot of the story of the scientific revolution also has a fictive quality. Notwithstanding all the ferment that was going on prior to the sixteenth century, historians and scientists often give the story a clear beginning, such as the publication of Copernicus's *De Revolutionibus Orbium*. In the manner of classic Aristotelian drama, the story has a phase of rising action (Brahe, Kepler), a climax (Galileo and his confrontation with the Catholic church hierarchy), and falling action (the codification of Western science by Bacon, Boyle, and Descartes). There is even a subplot: the story of William Harvey and the circulation of blood or Boyle's experimental demonstrations with the air pump. The ending varies a great deal, but often the last event in the story is Newton's *Principia*.

The story has been told and retold so many times that it has, to some extent, obtained some of the characteristics of a legend or myth. As with a legend, the story has a historical basis, but the protagonists have achieved a heroic, larger-than-life quality. As with many myths, the story has an etiological quality. In other words, it tells us how something came to be: it is about the origin of modern science and, to some extent, modern society. Like myths and legends, the story is no longer associated with a specific author but has, instead, passed into oral and popular tradition where, like a smooth pebble in a stream, it has become rounded over by the waters of popular consciousness. Rough edges—the historical details that do not fit—have been polished out of the story. What emerges in their place is a narrative of the triumph of reason, which, like a torch passed on from one runner to another in a relay race, is passed on from one heroic Great White Man to another.

Why tell the story one more time? Any understanding of modern science, technology, and culture requires a definition of the term *Western science*. Arguably a good way to begin to understand—and to interpret critically—the idea of *Western science* is to return to the origin tale. Keep in mind that there are many ways of telling the story. I could begin earlier, end later, add or cut characters, emphasize industry, deemphasize religion, make minor figures important and major figures unimportant, open up the story to the Muslims and other non-Western peoples, close the story in on England and Western Europe, and so on. The possibilities are endless. I give here something approximating the plot summary of what we might call the standard, Western textbook version of the story of the scientific revolution.

In 1543 Copernicus published *De Revolutionibus Orbium*, in which he argued for a sun-centered theory of the universe. Copernicus's main achievement was to elaborate the heliocentric model mathematically; he introduced no new data and in fact relied on Ptolemy's data. Because Copernicus, like Ptolemy, assumed perfect circular orbits, he still had to use epicycles. Even though Copernicus's predictions were no better than those of Ptolemy, Copernicus's theory has generally been hailed as mathematically simpler and more elegant.

The transition to the Copernican system was a gradual one. One impetus for the transitions occurred in 1577, when a comet appeared and cut across the various celestial spheres that were supposed to be the provinces of the various planets. The Danish astronomer Tycho Brahe also developed a huge pool of new data, and he supported a compromise system in which some planets move around the sun while the sun and its satellites move around the earth. When examining Brahe's data on the motion of Mars, his student Johann Kepler came up with a theory of elliptical motion, and he proposed a series of laws of planetary motion. The transition from circles to ellipses made it much easier to accept the Copernican theory.

Galileo provided additional support for the Copernican theory when he trained his new telescope on Jupiter's moons, which revolved around the planet and therefore provided a suggestion of how the planets might revolve around the sun. He also showed how the moon had mountains and "seas," and was, therefore, far from the perfect celestial body described by Aristotle. Galileo published his results in 1610 in the book *The Starry Messenger*, which created a great stir. However, what caused his famous trial and imprisonment was the book he published in 1632, *Dialogue Concerning the Two Chief Systems of the World*. In the book he provided a detailed criticism of the fundamentals of the Aristotelian science of motion, and he introduced experiments to test many of Aristotle's ideas. For example, Galileo showed through experiment (sometimes he is said to have dropped a weight from the tower of Pisa) that both heavy and light bodies accelerated at the same rate, the opposite of what Aristotle had argued. Moreover, Galileo asserted that the movement of celestial and terrestrial bodies followed the same set of physical laws.

The great problem of the Copernican system was that without crystalline orbs to hold the planets in place, nothing seemed to prevent them from crashing into each other or falling into the sun. Isaac Newton suggested an answer in his book the *Principia*, which was published in 1687. Newton owes some debt to the work of William Gilbert, who in 1600 published a book on the magnet and suggested that the earth was itself a giant magnet. That idea became one of the sources for the theory of a gravitational force that could hold planets in their orbit. Gravity was required to balance out the opposing centrifugal force that would otherwise tend to make the planets fly off into outer space like rocks from a sling. Newton argued that the force of gravity acting on falling terrestrial bodies was roughly equivalent to the force required to keep a celestial body such as the moon in its orbital path.

The foundations of other branches of the modern sciences were also laid during this period. In chemistry Robert Boyle distinguished the element and the compound, defined chemical reaction and analysis, and developed the law stating that the volume of a confined gas at a constant temperature decreases with pressure. In medicine and biology in the middle and later decades of the sixteenth century, the Italians Vesalius, Colombo, and Fabricus took steps toward the theory of the circulation of blood, beginning with the idea that the blood does not cross the septum but instead circulates from the right side of the heart through the lungs and back to the left ventricle. The Englishman William Harvey studied at Padua and by the 1620s had developed his predecessors' ideas into a comprehensive theory that the blood circulates from the heart through the arteries and back through the veins to the heart.

In the early seventeenth century Bacon and Descartes provided a philosophical or methodological rationale for the growing movement of natural philosophers. Notwithstanding the differences discussed in the previous chapter, their work helped legitimate the subsequent founding of the first scientific societies, most notably the Royal Society of London in 1660 and the Académie Française in 1666. The two philosophers provided the first formulations of the scientific method, a term granted such status in the story that we might do better to capitalize it as the Scientific Method. In the past, historians and philosophers have argued that the Scientific Method eventually became a combination of deduction and empirical observation, a union of Descartes and Bacon. More recently, studies of scientists in practice reveal that they do not adhere to a single Scientific Method or a even a single set of scientific norms. Instead, they pursue their research in a flexible and opportunistic way. Thus, the divergence between Bacon and Descartes could be used to argue that the so-called Scientific Method was, even in its origins, more of a plurality of—and back to the small letters—scientific methods.

The Origin Story: Some Multicultural Questions

A conventional narrative about the scientific revolution has several common features. First, it tends to tell the story in intellectual terms. The social is filtered out, except when it is useful to construct science as a triumph over superstition. Society figures in the background as a source of secondary characters: enlightened princes who support science and dogmatic church leaders who oppose it. Second, the story focuses on Europe, and it is told in the form of a dialogue between the Old Europe (ancient thinkers) and the New Europe (modern thinkers). What goes on during the time between the Old and New Europe, not to mention what happens before the Old Europe, is basically put into a black box. Exchanges between the West and the rest are written out of the narrative or relegated to secondary plot status. Third, the story is told as an event. A scientific revolution occurred. It is a dramatic change, not a gradual evolution.

Historians have long quarreled with each of the assumptions of the conventional narrative. Yet, the narrative continues to appear, especially in popular forums such as television programs. Told in conventional form, the narrative of the scientific revolution contributes to the ongoing ethnocentrism of West is best. In this sense the narrative can be seen as colonialist.

Let me begin my critique of the conventional scientific revolution narrative with a question: "How Western is 'Western' science and how revolutionary was the scientific 'revolution'?" One starting point for the answer to this question is the work of the scientist, historian, and sinologist Joseph Needham. Needham's studies of science in China were framed around the question, "Why did the scientific revolution not take place in China?" Although that question may seem to serve the West-is-best ideology, Needham documented the scientific and technical achievements of China, and in the process he helped erode the idea that other cultures are incapable of producing sophisticated scientific research.

Needham has also contributed to a critique of Western ethnocentrism by documenting the tremendous interchange between European and non-European cultures prior to the scientific revolution. He has argued that China was generally more advanced technologically than the West until about the sixteenth century. He has also shown that in many cases it is very likely that inventions and discoveries were made first in China and then transferred to Europe. Such innovations include, as Needham enumerates in one of his lists, magnetic science, equatorial celestial coordinates and the equatorial mounting of observational astronomical instruments, quantitative cartography, the technology of cast iron, essential components of the reciprocating steam-engine such as the double-acting principle and the standard interconversion of rotary and longitudinal motion, the mechanical clock, the boot stirrup and the efficient equine harnesses, to say nothing of gunpowder and all that followed therefrom. (1974:116)

Much of the technical infrastructure of what is considered to be Western science—including much of its mathematics—rests on borrowings from China and other non-Western cultures.

Other historians have begun to question the scientific revolution in a more profound way. They have increasingly looked back to earlier centuries and developed a more gradual narrative that shows ongoing developments in the centuries prior to the fifteenth century. Furthermore, by looking more carefully at medieval science and its achievements, historians are able to open up the complicated question of how Western science was deeply dependent on complex interchanges with non-Western sciences. As the historian Saleh Omar notes, extending the scientific revolution back to the thirteenth century

gives fresh vigor to the question of Arab influence on medieval Latin science and, ultimately, on seventeenth-century European science. The revival of scientific activity in Europe in the thirteenth century followed the translation into Latin of many Arabic works on optics, astronomy, mathematics, and medicine, as well as many of the Greek works, which were also translated from Arabic, toward the end of the twelfth century and the beginning of the thirteenth. (1979:68)

The conventional Western view of Arabic-Muslim science has tended to overlook its originality and its influence on the scientific revolution in Europe. For example, in a 1931 essay that reviewed the translation and transmission of Arabic-Muslim texts, the historian Max Meyerhof invoked the word *storehouse* to describe Arabic-Muslim science (Meyerhof 1931:344). Summarizing its role, he wrote:

Looking back we may say that Islamic medicine and science reflected the light of the Hellenic sun, when its day had fled, and that they shone like a moon, illuminating the darkest night of the European Middle Ages; that some bright stars lent their own light, and that moon and stars alike faded at the dawn of a new day—the Renaissance. (354)

Although an older generation of scholars recognized the achievements of Arabic-Muslim and other non-Western scientific traditions, many of those scholars also invoked the storehouse image and downplayed the direct influence of Arabic-Muslim science on the leaders of the "scientific revolution." One exception is the historian and medievalist Herbert Butterfield, who in *The Origins of Modern Science* noted that a thirteenth-century Muslim physician argued long before Harvey that the only way that the blood could go from the right to left ventricle of the heart was through the lungs (Butterfield 1957:55). Although Harvey's discovery of the circulation of the blood may be a case of rediscovery, the rediscovery itself was in part shaped by the Arabic-Muslim scientific tradition. The Arab philosopher Averroës (or Ibn Rushd) had a great influence on the university at Padua, where Harvey studied. In turn, Harvey's work owes a great deal to his teachers and predecessors at Padua, who nurtured experimental approaches in biology. Although Padua was an Aristotelian school, the university approached Aristotle from a relatively secular and therefore critical viewpoint. As Butterfield comments, "The Paduans were inclined to adopt this attitude because they were an Averroist university—seeing Aristotle in the light of the Arabian commentator, Averroës" (60).

Research on the Arabic-Muslim influence on early modern physics and optics has taken great strides in recent years. The historian George Saliba (1987) has put to rest the ethnocentric idea that Arabic-Muslim science served only as a storehouse of Greek science that was activated during the Renaissance. Saliba and other scholars have argued instead for a "scientific revolution" in Arabic-Muslim thought that took place during the thirteenth and fourteenth centuries and has become known as the Maraba school. Historians have demonstrated close parallels between Copernicus's work and that of the Arab astronomer Ibn al-Shâtir of Damascus. The details of the transmission remain to be worked out, but it is now clear that Copernicus borrowed heavily from the developments achieved in the Arabic-Muslim school (Saliba 1991).

A similar argument may be possible to make for Galileo, who, like Harvey, was a student at Padua. Although by the 1620s the university had earned a reputation as a conservative place where peripatetic scholars defended Aristotelian physics against the new ideas of Galileo, Butterfield argues that Galileo himself may have been influenced by the general experimental approach to biology and the secular approach to Aristotle. It is also known that Galileo had a copy of the *Opticae Thesaurus* of Ibn al-Haytham (Alhazen), an Arab scientist who was known for his experimental method (Omar 1979:68). Galileo used Ibn al-Haytham's work to refute the theory that the moon was a polished mirror, and it is likely that the Muslim scientist may have had a more general impact on Galileo. Furthermore, Kepler also studied Alhazen's works, and the Arab optics specialist influenced a number of other Western scholars as well (ibid.).

The influence of Arabic-Muslim science on the leaders of the scientific revolution has clearly become a topic of serious scholarship. Similar work shows how a number of technological innovations in the West— weight-driven and water clocks, glass making, ogival (Gothic) architecture, water-raising machines, and paper making—were technology transfers from the Islamic world (Al-Hassan and Hill 1986:31–35). In turn, studies of the transmission of medieval Islamic science and technology to the West may lead to the recognition of other, more complex patterns of transmission.

One area of scholarship that is opening up the question of a multicultural history of science and technology is the history of mathematics. For example, in *The Crest of the Peacock* the historian of mathematics George Gheverghese Joseph argues that Eurocentric accounts tend to begin with classical Greek mathematics and to minimize the influence of Egyptian and Mesopotamian mathematics on the Greeks. Joseph shows how the pre-Greek mathematics of the Egyptians and Mesopotamians was considerably more sophisticated than previously understood, and those mathematical traditions may also have had more influence on the Greeks than previously recognized. He also opens the "black box"—or "multicolored box"—of the European Dark Ages to reveal a complex set of ongoing exchanges among Arab, Indian, Chinese, and other non-European mathematicians. For example, Joseph shows how the medieval Indian mathematicians in Kerala developed mathematics close to the calculus, and he opens up the still unanswered question of Indian influence on medieval Arab mathematicians and, through them, on modern European mathematics. Joseph also debunks the storehouse theory of medieval Islamic science by explaining the crucial role that Islamic mathematicians played in bringing the geometrical traditions of the Greeks together with the algebraic and arithmetic approaches of Babylonia, India, and China.

The careful work of scholars of medieval Arabic-Muslim science is also undermining the extent to which the scientific revolution was revolutionary. The very idea of the Scientific Revolution may someday come to be rejected as ethnocentric. Although the European natural philosophers of that period certainly deserve credit for their achievements, there is an ideological quality to a way of writing history around a great watershed, before which is prescience and after which is science. There were other revolutions, or dramatic shifts, of scientific and technical knowledge, especially outside the West, such as the Maraba school. To the extent that other developments are not given the label of *the scientific revolution,* the idea of a single scientific revolution tends to put the Great White Men of early modern Europe on a pedestal, and to relegate everyone else to a secondary status.

One of the implications of opening up the black box of the sources of the "scientific revolution" is that the very idea of "Western science" is becoming increasingly problematic and ethnocentric. Certainly it seems better to talk about modern or cosmopolitan science rather than Western science, because science is now practiced throughout the world just as it was during the period prior to the scientific revolution. What may be happening is that Europeans and European-descent populations are beginning to recognize that the tremendous international exchange that characterizes cosmopolitan science and technology today may not be unique to the nineteenth and twentieth centuries. Instead, even during the period of the "origins" of modern science, and before it, there were ongoing exchanges across Asia, Africa, and Europe. Unlike political institutions and forms of social organization, which travel slowly, scientific and technical ideas diffuse rapidly across cultural boundaries. They are carried around in people's heads and can often be adopted without threatening existing power structures. Thus, science and technology tend to travel lightly and quickly.

Thinking of science from a multicultural perspective can be distinguished from thinking about it from an international perspective. Although historical research is showing increasing evidence for connections among European and non-European mathematics, technology, and natural knowledge, a case can be made that different cultures still maintained their distinctive styles of scientific thought. The case of Greek geometry versus Near Eastern and Indian algebra is one example. In this sense the different mathematical and scientific traditions constitute streams of ethnomathematics and ethnoscience. The science of Copernicus, Kepler, Harvey, Bacon, Galileo, Descartes, and Newton might better be thought of not as a development of a transcendent international science but as a specifically early modern Western elaboration and development of a multicultural scientific and mathematical legacy.

Technototems in Early Modern Science

I prefer to think of the European natural philosophers from Copernicus to Newton as thinkers who were reinterpreting their natural world in light of changes going on in their social world. They were producing a particularly Western spin on an international scientific legacy, and their ideas were coming not from transcendental genius but from the general debates and discussions going on in their contemporary societies. Thus, it can be a useful exercise to compare the early modern *natural* philosophers of Europe to the early modern *social* philosophers such as Machiavelli, Hobbes, and Locke.

The natural and social philosophers shared a number of common ideas that suggest the ways in which science was participating in the production and reproduction of early modern society. For example, the "Western" style of constructing science reveals a fascination with the idea of natural laws that govern the circulation and mechanical motion of bodies: celestial bodies, terrestrial bodies, animal bodies, economic bodies, social bodies. Furthermore, the emergence of an analytical and empirical approach to nature that broke it down into corpuscles, elements, and other small units is similar to the ideology of individualism in theories of society. John Locke and other early modern bourgeois philosophers argued for a view of the state as a social contract among consenting individuals, with the laws of the bourgeois state governing the actions of individuals much as God's natural laws governed the actions of individual planets and atoms. Furthermore, the elaboration of methodology as a kind of charter for scientific research ends up providing science with a body of rules, much in the same way that a constitution provides the state with a body of rules or the bylaws provide a company with rules of operation. Although the scientific method, constitution, or bylaws may not be followed in practice (and are, therefore, not necessarily norms), they become points of reference in an ideology that is constructed around the practices of science, the state, and the firm. Finally, the construction of experimentation as a public forum was similar to the emergence of voting and elections as rationalized ways of resolving public disputes or markets as a rationalized way of resolving differences between supply and demand.

The science of the scientific revolution is therefore an "ethnoscience" because it articulates and helps constitute the culture and society in which it is produced. On the one hand the science of sixteenth and seventeenth century Europe borrows on the knowledges of many cultures, and in this sense it is a multicultural science. Yet, on the other hand the elements of diverse knowledge systems are woven together in a way that is shaped by and reproduces the culture of the societies of early modern Europe, particularly the emergent bourgeois cultures of Northern and Western Europe.

Many of the studies of the science/society genre have focused on the question of how society shaped early modern science. Those studies focus on a set of factors that served as either spurs or inhibitors to the development of science. While interesting, those discussions can easily fall into antiquarian debates that have little general interest. I study the natural philosophy of early modern Europe because it put into play a number of assumptions about science and the natural world that inform the ways in which science and technology are carried out today. At the same time, the societies of Europe were developing new institutions and ideologies that are also in place today: capitalism, constitutional governments, religious pluralism, etc. As the historians Steven Shapin and Simon Schaffer have argued in *Leviathan and the Air Pump*, natural philosophy played a role in the production of these modern societies just as science today is part of their reproduction. The following sections therefore will examine how the making of early modern science was part of the making of early modern society. I will focus on four basic areas: Protestantism and capitalism, naturalistic orthodoxy, patriarchy, and colonialism.

Protestantism and Capitalism

Sociologists of science have argued that one key aspect of the nascent society was that social conditions allowed science to emerge in an institutionalized form as an *autonomous* social activity. In Western Europe a general historical process led to the separation or differentiation of various spheres of society—the economy, state, church, and so on—which also allowed for the development of science as a relatively autonomous activity, that is, as relatively independent of direct religious or political influence. That process culminated in the formation of scientific societies in the seventeenth century, but, as the historian Toby Huff notes in his comparative study *The Rise of Early Modern Science*, from the twelfth and thirteenth centuries crucial developments in the rationalization of law and religion in Europe lay the intellectual groundwork for the scientific revolution.

The concepts of autonomy, rationalization, and differentiation as keys to the idea of modern society owe a great deal to the work of Max Weber (1978). The German sociologist is probably the single thinker most associated with the problem of modernity and why it occurred in the West. He used the term *rationalization* to describe the general process by which societies are increasingly governed by more complicated bureaucracies, increased differentiation of functions, and ever more elaborate systems of rules and procedures.

In Western countries the process of rationalization underwent dramatic shifts during the period from approximately 1500 to 1800, or the period generally known as the emergence of modernity. In the political sphere, for example, a number of revolutions resulted in the transformation of political systems from monarchies, in which authority rested on a rule of tradition, such as hereditary succession within a royal family, to constitutional monarchies or republics, in which authority rested on a claim to represent society and a rule of law. Likewise, in the economic sphere the aristocracy, whose wealth had been measured in terms of royal privilege and ownership of land, was superseded by the emerging urban capitalist class, whose wealth rested on the accumulation of capital through industrial production and trade through markets.

The emergence of Protestantism and modern science were two intellectual developments that went hand in hand with these broader shifts in the social structure. Karl Marx and Frederick Engels noticed the connection, but it was not until Max Weber's classic study, *The Protestant Ethic and the Spirit of Capitalism*, that a sustained sociological analysis of the problem appeared. Weber began with the insight that in the Europe of his day, and particularly in Germany, Protestants tended to be disproportionately represented among the owners of capital, leaders of business, and higher levels of skilled workers. He also noted that the areas where capitalism was most advanced (e.g., the Netherlands, England, New England, and Scotland) were also areas where Protestantism flourished. As Weber himself recognized, and subsequent scholars have also argued, advanced industrial capitalism was by no means restricted to Protestants and Protestant regions. For example, the sociologist Robert Bellah (1957) argued that the religion of Tokugawa Japan was in some ways comparable to Protestantism. However, Weber found that the relationship between Protestantism and capitalism was pronounced enough in Europe to warrant further investigation and analysis.

Weber argued that certain aspects of Protestantism were conducive to a rationalized form of industrial capitalism that had emerged most prominently in Western Europe. The question of which way the lines of causality ran (capitalists chose Protestantism because it matched their interests, or Protestants became good capitalists because they were hardworking and nonfrivolous, or both) is entirely debatable—and it has been debated in a seemingly endless sociological dialogue. My own reading of the debates has led me to conclude that capitalism was generally the determining factor (capitalists chose and shaped Protestantism), although there is room for considerable local variation. However, I prefer to avoid the chicken-and-egg question altogether because it tends to lose the broader point that capitalism and Protestantism were two domains of early modern culture that had a curious convergence of structure. Borrowing an alchemical metaphor from the poet Goethe, Weber said they had an "elective affinity" for each other. I go into Weber's analysis in some detail because the relationship between Protestantism and capitalism provides a model for one type of analysis of the relationship between science and capitalism.

Weber argued that the earlier form of Protestantism—Lutheranism, which appeared in northern Germany and Scandinavia—had developed a key idea that was conducive to the methodical pursuit of wealth necessary for the modern form of capitalism. That idea, "the calling," implied a sense of personal commitment to work that is conveyed in the term *vocation,* or, in today's parlance, *career.* Although professionalization of most occupations, including science, did not occur on a large scale until the nineteenth and early twentieth centuries, the idea of a vocation was arguably one ingredient in the establishment of scientific inquiry as a relatively autonomous social domain.

Weber also argued that Lutheranism was only a first step in the development of a form of Protestantism that was conducive to capitalist activity. The second step, Calvinism, was associated with the Swiss theologian Jean Calvin but was also influential among the Puritans in England, Scotland, and the British colonies of the New World. In Calvinist theology salvation could no longer be mediated through the church and the sacraments; instead, it depended on the believer's inner state and level of faith. Catholicism did not require the same degree of "internalization" of faith—one could be a good Catholic by attending church, making contributions, receiving the sacraments, and doing "good works." For the Calvinist, however, salvation depended on one's inner state of grace.

The Calvinist doctrine, however, leads to an ironic twist: to be sure of one's state of grace, one engages in worldly activities—service to the community and hard work—that are dedicated to the glory of God. Good works are, in Weber's terms, "the technical means, not of purchasing salvation, but of getting rid of the fear of damnation" (1958:115). Furthermore, not only must one do good works, but one must avoid doing bad works; in other words, the Calvinist doctrine calls for a high degree of self-discipline. The life of methodical self-restraint provides the perfect formula for the capitalist: someone who works hard and spends little. The Calvinist God watches over the faithful all the time. Like Santa Claus, who knows who is naughty or nice, when people are sleeping and awake, the Calvinist God keeps track of who has been good or bad. In a sense, he keeps a ledger on everyone. God is an accountant of sin, much as the Calvinist capitalist is an accountant of funds.

Three of the most influential aspects of the modern Western cultures of Protestant Northern Europe and North America are expressed in the Calvinist formulation of Protestantism. The first aspect, individualism and self-reliance, meant relying only on one's own faith and the scriptures, not on the authority of the church or state. The second, good works and community service, became transmuted into patriotism, civic heroism, and the civil religion. The third, the work ethic, called upon the faithful to engage in worldly activity such as business enterprises, and it even transformed the accumulation of wealth into a badge of God's grace. As the popular saying goes, God helps those who help themselves. At the same time, if the faithful were to enjoy the wealth they worked so hard to accumulate, they would risk falling from the straight and narrow, and therefore succumbing to the pleasures of the flesh and the temptations of the "evil one." Thus, as Weber put it, "When the limitation of consumption is combined with this release of acquisitive activity, the inevitably practical result is obvious: accumulation of capital through ascetic compulsion to save" (1958:172).

Weber was not extraordinarily optimistic about the prognosis of modern civilization. He believed that the world was becoming increasingly rationalized, bureaucratized, and materialistic. In short, the world had lost touch with the original religious sense of a calling and the glorification of God through one's work. At the end of *The Protestant Ethic* Weber described a Frankensteinian society in which industrial capitalism was so out of control that it had reached the point of threatening to swallow up civilization and humanity itself. His lament is a familiar one repeated by many social thinkers, so much so that the culture critic and political theorist Langdon Winner (1977) has given it a label: the "technics out of control" theme. In a frequently cited passage Weber noted, "The Puritan wanted to work in a calling; we are forced to do so" (1958:182). He saw the end result of ascetic capitalism as the great industrial society that has grown out of control. For the Puritan, concern with material goods was supposed to lie on the shoulders "like a light cloak," but, as Weber wrote, "fate has decreed that the cloak should become an iron cage" (ibid.).

Weber had much less to say about science and technology per se, but he clearly believed that they were part of the same general historical process of rationalization and modernization. In a footnote to *The Protestant Ethic* he wrote that, for the Puritans and other ascetic Protestants,

> just as the Christian is known by the fruits of his belief, the knowledge of God and His designs can only be attained through a knowledge of His works. . . . The empiricism of the seventeenth century was the means for asceticism to seek God in nature. It seemed to lead to God, philosophical speculation away from Him. (1958:249, n. 145)

I would add to Weber's footnote that Protestants rejected the role of church dogma as the intermediary between the faithful and God, and instead they insisted on authority anchored in a personal reading of the scriptures. Luther translated the Bible into German so that his followers could read it and study it for themselves, and the Protestants in other countries soon undertook similar projects. Thus, in the area of religion Protestants were opening doors to what one might call a more empirical approach to the scriptures, that is, one based on reading them directly instead of obeying the interpretations made by a priestly hierarchy.

Regarding the questions of civil religion and the work ethic, it is also possible to draw out the implications of Weber's analysis of Protestantism for science. For example, Francis Bacon argued for the value of natural philosophy because it would make natural knowledge available not only for the inquirer but also for the state and society. Likewise, the work ethic oriented action toward this world, and it made physical labor a sign of honor. In a similar way, the emerging natural philosophies focused on this world, and they made the physical labor of experimentation into a centerpiece of their methods. The *labor*atory, then, had a certain elective affinity to the work ethic.

Robert Merton, one of the founders of the sociology of science in the United States, developed Weber's thesis in his doctoral dissertation, which was subsequently published in 1938 as *Science, Technology, and Society in Seventeenth-Century England*. Merton argued that the Puritans' emphasis on reason, "based partly on the conception of rationality as a curbing device of the passions," was conducive to the scientific mindset (1970:92; see also 66–68). Merton also extended Weber's comments on empiricism and Protestantism by linking the experimental method to Puritanism's utilitarian emphasis:

> The Puritan insistence upon empiricism, upon the experimental approach, was intimately connected with the identification of contemplation with idleness, of the expenditure of physical energy and the handling of material objects with industry. Experiment was the scientific expression of the practical, active, and methodical bents of the Puritan. This is not to say, of course, that experiment was derived in any sense from Puritanism. But it serves to account for the ardent support of the new experimental science by those who had their eyes turned toward the other world and their feet firmly planted on this. (93)

Merton adds that although Puritanism may not have had a direct effect on the development of the scientific method, it helped legitimate science by constructing it as a noble activity and profession (94–95).

The historian Gary Deason (1986) developed this argument by showing parallels among the ideas of nature held by natural philosophers and Reformation theologians. Calvin, for example, saw the natural world as made up of merely the "instruments" of God's creation. In other words, he stripped the natural world of its animism and helped turn it into a more inert world of objects. The philosophers of the scientific revolution, such as Robert Boyle, developed this view of the natural world in their own thought and writings. By seeing nature as inanimate, it became easier to think of it as being subjected to lawful patterns and regularities. God entered into the picture as the cause behind motion, or, in Newton's theory, as the force behind gravity.

In *The Cultural Meaning of the Scientific Revolution* the historian Margaret Jacob has further explored the ways in which the new natural philosophies served as totems that became identified with various types of Protestantism and antipapalism. Some of the apologists of the emerging fields of modern science, such as Francis Bacon, explicitly criticized the Catholic church and aligned the new natural philosophy with the Anglican church. Bacon believed that science would help usher in a new era—a new millennium—and this millenarian impulse was also shared with Bacon's Puritan contemporaries (Jacob 1988:32–36). Furthermore, by the 1650s the belief that science could help transform society had been adopted by the Anglican orthodoxy, and therefore the new science became increasingly aligned with the "liberal" Anglican position that tolerated the Puritan sects. Indeed, Jacob argues that by the time of Newton science was less an expression of Puritanism than of the liberal or "latitudinarian" Protestantism typifying the Anglican church of the late seventeenth century (87–98).

Historical studies of early modern Europe therefore point to a number of ways in which Protestantism, capitalism, and the new natural philosophies became interwoven domains of a new type of society. The connections are by no means uniform across the various times and countries of early modern Europe. For example, the religious connection was less well defined in the sixteenth century than in the seventeenth. Protestants persecuted Kepler on the continent, and he had to take refuge with the Jesuits at the end of the sixteenth century. In England a number of Protestants supported a visit by the anti-Copernican Comenius (Kominsky) as well as publication of several editions of his books. On the other side Copernicus was a priest in the Catholic church, and he had support from his superiors for the publication of his work. However, by the 1620s and 1630s—and especially after Galileo's prosecution— lines were hardening in the Catholic church, and subsequently the natural philosophers found a more hospitable home in the Protestant countries, especially among the emerging merchant classes within those countries. For mid-seventeenth-century England there is substantial (but still contested) evidence in support of a Puritan connection. By the late seventeenth century and the time of the Newtonian synthesis, the connection had shifted more to a liberal form of Anglicanism. Thus, to speak of a Protestant connection is to speak of specific forms of Protestantism that varied across times and national cultures, and also to recognize that not all Protestants everywhere supported the new natural philosophies. Likewise, not all the new natural philosophers were Puritans or even Protestants, nor did they live in Protestant countries. It is more that both Protestantism and science were expressions of the emerging modern capitalist societies in Europe, and the two tended to share some fundamental ideas with the new social forms of statehood, economy, and society.

Just as Protestantism was closely connected to the emerging classes that eventually became today's capitalists, so the new natural philosophers frequently found a welcome home among that social class. Boris Hessen, a Soviet physicist and historian who was a contemporary of Merton, developed a good case for a direct connection between science and capitalism that did not require Protestantism as an intermediary. He argued that most of the solutions and theories in Newton's *Principia* are closely tied to problems his contemporaries faced in the areas of water transportation, mining, and warfare. Along the same lines, Jacob has also shown how quickly the natural philosophers attempted to apply their knowledge in the service of the state, the merchants, and eventually the new industrial class. I would add that in Italy Galileo was quick to make a profit by selling the telescope to merchants, and his compatriots were also demolishing Aristotle's theory of motion by showing how the new theory of canonball trajectories would lead to more accurate warfare. Thus, whether one wishes to trace out affinities among the natural philosophies, Protestantism, and capitalism, or simply go directly to the instrumental basis of many of the developments in natural philosophy, the "bottom line" remains: the new natural philosophies were deeply interwoven with the emergent modern/capitalist society.

Notwithstanding the differences between Weber-Merton and Hessen, both provide a framework for seeing early modern science in totemic relationship with an emerging bourgeois society. At a general level the early modern natural philosophers opposed the dogmatism of scholastic philosophies of nature just as the Protestant reformers opposed dogmatic religious theologies or capitalists opposed aristocratic privilege and restraints on market economies. Certainly, there was a great deal of room for local variation and for individuals whose ideas were not in step with their societies. Yet, there is also a large body of evidence that suggests specific relationships among theology, natural philosophy, and emergent social classes and institutions of the type outlined here. The systematic empiricism and mechanism that the early modern natural philosophers used to describe the natural world was one version of an overall cultural politics of rationalization that encompassed the state, the economy, religion, and other domains of life.

Naturalistic Orthodoxy

The early modern natural philosophers may appear to be the radicals and heretics of their time. The case of Galileo's persecution by the Catholic church contributes to that image, and there are numerous cases of persecution or attack by both the Catholic church and even by some Protestant clergy. What may be less well understood is the common ground that was constituted in the debates between the natural philosophers and church leaders. By the time a peaceful coexistence was established, as first occurred in the Protestant countries, the natural philosophers and church leaders had come to share a common enemy: witches, sorcerers, and an underground of other magical practices and occult groups. Although the churches had opposed those groups on moral grounds, the early natural philosophers began to develop a rationale for opposing them on philosophical grounds. In the process, heresy was being transformed into heterodoxy or pseudoscience.

Much of the work on the occult in early modern Europe has been pioneered by the historian Frances Yates, whose research has revealed a complex culture hidden beneath the surfaces of the official stories of science, church, and state. Renaissance thinkers showed a great deal of interest in various occult ideas such as magic, astrology, and alchemy (age-old ideas that today are sometimes called New Age ideas). Among the Renaissance doctrines was hermeticism, whose followers held that human beings were capable of sharing in divinity and even of communing with God directly. They also believed that the stars and planets have souls and can be operated on through magical spells to change individual lives and social destinies. The doctrine therefore can be thought of as parallel to astrology today, with a mystical angle that involved belief in direct divine knowledge through religious experience.

The term *hermeticism* derives from Hermes Trismegistus, whose writings were being translated during the Renaissance. Although he turned out to be a Greek philosopher from the third century C.E., at the time he sparked a great deal of interest because Renaissance scholars believed him to be an ancient Egyptian, perhaps contemporary with Moses. Thus, to the Renaissance scholars his wisdom represented the pure knowledge of the dawn of time. Hermeticists believed that the "ancient true philosophy" of Hermes Trismegistus had been corrupted as it was passed down through the Greeks and the Judeo-Christian tradition.

Hermetic doctrines played a role in many aspects of early modern science. For example, Galileo's defense of the heliocentric view of the universe and his persecution by the Catholic church are well known, but less well known are the ways in which Galileo had come to be identified with hermeticism. In 1584 the Italian intellectual and diplomat Giordono Bruno had written *La Cena de le Ceneri* (*The Ash Wednesday Supper*), in which he argued that the Copernican view signaled a return to the "ancient true philosophy" of hermeticism. Bruno also believed that a consequence of accepting the ancient philosophy would be the unity of liberal English Protestants and liberal French Catholics. The sun (including the sun-centered view of the universe) became a symbol of the political program of unification through the occult philosophy of hermeticism. In 1591 Bruno was captured by the Inquisition, whose officers placed him in prison and, after several years, burned him at the stake. Politically, Bruno's prosecution was a signal sent by the Pope, who was allied with the Hapsburgs of Spain and Germany, that no such deviations would be tolerated. In such a context heliocentric theories took on highly charged cultural meanings associated with treasonable politics.

Although Galileo was probably not very interested in hermetic ideas, he faced the stigma of being labeled a hermeticist. Other early natural philosophers were openly interested in hermeticism and alchemy. For example, in *The Rosicrucian Enlightenment* Yates traces hermetic or occultist influences up to the founding members of the British Royal Society. Newton devoted an enormous amount of time to the study of alchemy, and the closer one studies Newton, the less modern he looks. In general, his work on alchemy as well as that of the more radical thinkers and hermeticists provides a reminder that the seventeenth cen-

tury was a very different culture from that of the nineteenth and twentieth centuries.

Although Newton and some of his fellow natural philosophers were
interested in hermetic and other occult ideas, those ideas quickly came
to occupy a heterodox position. In their place the natural philosophers
came to accept what I will call a naturalistic orthodoxy, which admitted a
spiritual order in the universe but denied the existence of magical powers, the intervention of spirits or demons, astrological forces, and alchemical secret knowledge. The new cosmological order well described the
emerging social order. The universe was divided into spiritual, natural,
and social realms, just as the rapidly differentiating social world admitted
divisions of authority among church, the emerging scientific world, and
the state. The leading groups in the three domains converged in their
shared interest of weeding out magical practices and alternative popular
knowledge. Puritan reformers smashed icons of magic in their churches
and the natural philosophers advocated experimental methods to
replace mystical intuition and magic.

Along with the constitution of a naturalistic orthodoxy, a countercultural heterodoxy emerged. In England Puritan reformers and revolution-
aries found themselves outflanked by a whole series of radical groups—
the Levellers, Diggers, Quakers, Ranters, and so on—who demanded a
series of political changes. Their reforms ran the gamut of their society:
politics (one "man," one vote), economics (property redistribution), religion (end of church tithing, complete religious toleration), society (right
of women to preach and vote in congregations), and lifestyle (criticisms
of puritanical restrictions on smoking, drinking, and sexual freedom; see
Jacob 1988:77). The radical groups also had a different view of knowledge: they were more interested in hermeticism and they often supported an immanent rather than transcendent view of God. In other words,
they believed God was in all things, especially in nature.

One of the leading natural philosophers of mid-seventeenth century
England, Robert Boyle, was part of the group of moderate Puritan reformers who rejected the assault from the more radical groups. Boyle's corpuscular philosophy—an early modern version of the atomistic theory of
matter—might in retrospect appear to be an important step forward in
the development of modern chemistry and physics, but at the time it also
had meaning in the context of the conflict between the moderate Puritan
reformers and their radical rivals. Boyle believed that matter was made
up of inanimate atoms that collided with each other, and that those colli-

sions were—through divine will—organized into patterns or laws. Thus, his theory could be tested through experimental observations of lawful chemical reactions (Jacob 1988:81).

At the same time, however, Boyle's corpuscular philosophy had political implications. His mechanical view of matter was consistent with the orthodox religious view that supported a dualism of matter and spirit. In contrast the radicals' immanent or pantheistic view held that spirit was in matter. Jacob argues that the pantheistic view of nature corresponded with the radical political viewpoint that questioned the traditional authority of the church and state. If "God" or the spirit was everywhere, in nature and in human beings, then it would be possible to think of a church as a community of believers that did not need the support of the official church or of the state. Nor, for that matter, would it be necessary to support the official church through the tithe, the mandatory tax of ten percent of one's income (82).

Boyle's corpuscular philosophy, therefore, became meaningful in opposition to the radical pantheistic views that were linked with the radical democratic movements of the time. In addition, the inductive or experimental side of the new natural philosophies was also connected with these political conflicts (Jacob 1988:82). The radical groups still maintained that knowledge could come through magic, mystical experience, or direct revelation. Unlike the radical groups, the natural philosophers held that knowledge comes through detailed observation, hypothesis testing, and experimentation. Although the natural philosophers granted a role for the channeling of divine knowledge, that channeling was accomplished through the scripture rather than through the various forms of direct experience that were and still are found in the counterculture of religious, medical, and scientific heterodoxy.

Thus, even in the seventeenth century the emerging capitalist class and their associated religious and scientific ideologies were already being questioned by more radical groups. Those groups advocated many changes that were shared by radical movements of later centuries, such as the utopian socialist groups of the nineteenth century and the communist movements that achieved power in the twentieth century. In addition to calling for a redistribution of wealth, various radical groups questioned the materialistic and mechanistic view of nature in favor of a more pantheistic and holistic view.

The countercultural heterodoxy also included popular beliefs and practices associated with witches, sorcery, and "wise" healers. The persecution of witches was a complex phenomenon that varied across locales, but several studies have demonstrated connections with capitalist-generated class conflicts as well as with the modernization of patriarchy. As early as 1584 critics such as Reginald Scot were propagating naturalistic theories that described witches as afflicted by natural disorders such as melancholy, the forerunner of what is today called depression. As part of the general Protestant assault on the Catholic legacy of magic, the naturalistic theories could replace the supernatural theories of demonic intervention and therefore increase the distance between popular and elite interpretations. Naturalistic theories of witchcraft also constituted the first steps toward the transformation of moral problems into medical ones. Still, whether witches were invoking supernatural powers or merely victims of melancholy, they were portrayed as pathological. Thus, with respect to the popular traditions, naturalistic theories of witchcraft represented a continuation of their status as heterodoxy, but now transmuted from the moral heterodoxy of heresy to the naturalistic heterodoxy of illness and pseudoscience.

A paradox is beginning to emerge. The literature on the capitalist-Protestant connection shows that early modern natural philosophies are totems of an emergent modern society that is marked by religious rationalization, capitalist economic relations, and constitutional government. Inasmuch as the modern natural philosophies were part of a rejection of an older social and intellectual order, they carried with them a progressive political message. However, the modern social order was constituted through the construction of a new orthodoxy and a corresponding counterculture. The counterculture was somewhat different from the heresy of the older society, for the new society allowed some room for debate and open discussion of heterodox ideas. A society emerged in which it was possible to have a counterculture of heterodox political, social, religious, and scientific ideas. As hermeticism and other occultist doctrines were driven out of natural philosophy, they found a home in the emergent radical groups. The groups proposed alternative knowledges and social structures that stood in opposition to those supported by both natural philosophers and official churches in much the same way that the popular traditions of local witchcraft, sorcery, and healing did. As a result, a countercultural knowledge tradition emerged. The basic cultural structure remained in place in later centuries, from the Spiritualist movement and utopian communes of the nineteenth century to the hippies and New Age movement of the twentieth century.

Patriarchy

One of the demands of the radical English groups was the right of women to preach, and many of the witches who were persecuted during the witch crazes were women. The gendering of the counterculture/orthodoxy relationship suggests yet another way in which the modern natural philosophies were part of the production of a new social order. As the historian David Noble argues in *A World Without Women*, modern science emerged from the cloistered tradition of monastic scholasticism. The men-only aspect of medieval scholarship has continued to inform modern science and the technical professions, which even today remain occupations dominated by men. Furthermore, as many feminist scholars have noted, not only are science and technology dominated by men along the dimension of institutions and occupations, but they are also often dominated and biased by masculinist assumptions that shape content. Feminist scholarship has shown that the masculinist assumptions of modern science are evident even in its first formulations. Thus, the nature/science relations in the emergent natural philosophies came to articulate female/male relations in the emergent capitalist patriarchies.

Returning to Robert Boyle, the philosopher Elizabeth Potter (1989) has added gender to the already complex picture. To review the levels of analysis, remember first the question of the making of a modern or bourgeois society. On that point Shapin and Schaffer (1985) argued that Boyle's experimental method contributed to the demarcation of autonomous and public spaces that accompanied the various rationalizing transformations of modernity. In terms of the question of orthodoxies and heterodoxies, Jacob showed how Boyle's corpuscular theory was constructed in opposition to the pantheistic theories of the radical groups. Potter points to yet another meaning of Boyle's work: the gendered aspects of the corpuscular theory. She reviews the gains of women in the independent churches and their demonstrations outside Parliament during the 1640s. She contrasts the activism and activity of those women with the ideal of the bourgeois woman who was to stay at home and raise the children. Finally, she notes how Boyle's laboratory assistant Robert Hooke describes matter as a passive, inactive female principle that he explicitly genders as "mater" and contrasts with the male principle of motion. Thus, not only do the corpuscles take a pantheistic God out of nature but they also build a passive femininity into it.

The thesis that modernity and early modern science reconstructed

Mother Nature into a passive and inert thing is the subject of Carolyn Merchant's *The Death of Nature*. The historian and physicist argues that early natural philosophers such as Bacon elaborated an explicitly gendered view of the relationship between nature and science. For them science was a masculine activity in which the task was to watch, classify, order, and dominate a natural world that for millennia Western cultures had associated with the female. By reconstructing this world as mechanical and inert, the early modern natural philosophers strengthened the male/female opposition as one of activity and passivity.

Not only did Bacon construe science as a masculine project of mastery and domination over a nature that he wrote of as female, but he believed in an ideal society that was thoroughly patriarchal. In the *New Atlantis* Bacon envisioned a utopian world based on scientific discoveries that contained the first plans for a scientific research institute. The utopian society would be ruled by scientists not unlike Plato's philosopher-kings, and the scientists were, of course, men. Bacon had served King James I of England in a number of positions, one of which involved representing the state against women accused of witchcraft. Merchant details a number of antifeminist and misogynist aspects of royal policies toward women, and she also draws a parallel between the inquisition of witches, including their torture, and Bacon's attitude toward nature. She argues that Bacon viewed nature as unruly and in need of domination, not unlike the witches and women of his day.

Although Bacon's views of the nature-science relationship were gendered, there are some disagreements about the degree to which the gendering was violent. Whereas Merchant emphasizes Bacon's construction of the science-nature relationship as rapacious, STS researcher Evelyn Fox Keller finds in Bacon more a suitor and seducer who is interested in "a chaste and lawful marriage between mind and nature" (cited in Keller 1985:36). However, it is an unfortunate truth for Bacon's society, as for our society today, that rape and marriage are not mutually exclusive.

Whether it was rape or seduction, Bacon portrayed the science/nature relationship as gendered domination. "For you have but to follow and as it were hound nature in her wanderings," he wrote, "and you will be able when you like to lead and drive her afterward to the same place again" (cited in Merchant 1980:168). Likewise, Bacon described research in terms of "disclosing of the secrets of nature," such that scientists should have no scruples against "entering and penetrating into these

holes and corners, when the inquisition of truth is his whole object"
(ibid.). Bacon viewed science as a process of taming nature, of putting
her into constraints, much as a woman would be restricted to the world
of the home and child rearing:

> She [i.e., nature] is either free and follows her ordinary course of
> development as in the heavens, in the animal and vegetable cre-
> ation, and in the general array of the universe; or she is driven out
> of her ordinary course by the perverseness, insolence, and for-
> wardness of matter and violence of impediments, as in the case of
> monsters; or lastly, she is put in constraint, molded, and made as
> it were new by art and the hand of man; as in things artificial. (cited
> in Merchant 1980:170)

As for other aspects of the new natural philosophy, the gendered relationship between science and nature became one of the deeper structures of modern scientific culture. The anthropologist Sharon Traweek
(1992b) has shown that even today physics is a highly gendered enterprise, in which the nature/science relationship reminiscent of Bacon is
still very much in effect. It is more than coincidence, for example, that
physicists find "charm" in elementary particles of nature (or, to invoke
another marker of Otherness, they also find "color"). More generally, distinctions between the natural sciences (masculine) and human sciences
(relatively feminine) play themselves out in metaphors such as "hard"
and "soft" sciences. Likewise, as I have shown in *Science in the New Age*,
ongoing debates today between skeptical materialists and New Age
advocates take place in a network of gendered metaphors. Although the
patriarchal gendering of nature is much older than natural philosophers
such as Bacon, they helped confirm and give new life to a gendered
world of metaphors and social relations.

Colonialism

Bacon also provides a point of departure for a cultural analysis of science and colonialism. For example, when he speaks of "binding" nature
and making her into a slave so that she could perform useful services, his
metaphors slip into those that are more readily associated with colonialism. Merchant rightly describes how Bacon's *New Atlantis* is a femi-

nist dystopia that relegates women to extraordinarily subordinate positions. I would add to her reading that the book is also framed in terms of a colonial expedition that sets off across the South Pacific from Peru. This point brings up another side to the construction of the science/nature relationship: its reproduction of the West/rest relationship.

The connection between science and colonialism is now well recognized. By the middle of the seventeenth century colonial powers were beginning to finance scientific studies as part of their process of conquest and colonization. For example, when the Dutch annexed the former Portuguese colony of Brazil in the 1640s, the colonial Dutch administrator sponsored studies of tropical diseases, flora, fauna, geography, astronomy, meteorology, and local life and culture. Likewise, in 1736 the French Académie des Sciences launched the first major scientific expedition to the Amazon, one that was dedicated to scientific study but brought back news of rubber, quinine, curare, and other products. The botanical gardens of Britain and other countries served as waystations for the transfer of crops among colonial lands and between those lands and the center. The transfer of rubber from the Amazon to Asian plantations, with its disastrous effects on the Amazonian economy of the early twentieth century, is probably the best known case. By the beginning of the twentieth century, science had come of age as an important ingredient in the project of European colonialism. Today, biotechnology laboratories have replaced botanical gardens, but the bioengineered food products of First World laboratories threaten to continue displacing tropical agriculture in favor of production sites that are best suitable for the profits of center countries.

Thus, Bacon's imagery of control, slavery, and harnessing of nature extends beyond the relations between white women and white men in Europe to the relations between Europeans and non-Europeans. As the historian Michael Adas shows in *Machines as the Measure of Men*, the new knowledge of Western science came to be used as a yardstick against which non-Western civilizations were measured. By the late seventeenth century European travelers were increasingly scornful of what they saw as the less advanced knowledge of non-Western peoples, even that of India and China. Those knowledges (and societies) came to be equated with the spurned knowledges (and politics) of countercultural groups within the center countries. Europeans could point to their superior technologies, particularly in the areas of warfare and navigation, as justifications for the civilizing mission of the "white man's burden." The project of imagining science as domination over nature consequently came to play a role in the related project of colonial domination over non-Western peoples, especially the "nature people" of indigenous societies.

Conclusions

The standard or popular version of the story of the origins of modern science still circulates through schools, media, and scientific communities. Its circulation carries with it a politics that contributes to the reproduction of a body of metaphors as well as a social order. Questioning the idea of a scientific revolution and its roots in the genius of Great White Men also means reinterpreting the technototems that are put into play in the story. I have drawn on a long and distinguished trajectory of scholarship that helps open up the standard version to critical, alternative readings.

Modern science and technology did not cause the various aspects of modern society: constitutional democracies, legal rationalization, social mobility, this-worldly asceticism, capitalism, patriarchy, countercultural heterodoxy, and racist colonialism. Nor did that social order create modern science and technology. Rather, science was one manifestation of the social order and one factor in its legitimation and production. Those who wish to change the culture can choose any number of sites as their point of entry. Wherever they begin, they will need to tell a new story about how they want things to be different. In the process they will be reconstructing their societies and its stories. What I have attempted to show here is an example of how that reconstruction works.

4 | Temporal Cultures and Technoscience

After the seventeenth century scientific inquiry became so diverse, multidisciplinary, technical, and international that it is almost impossible to achieve an understanding of anything more than general outlines or very specific cases. In this chapter, as in the previous one, I examine the cultural politics of the modern history of science. I will continue to develop the argument that science is an integral part of the general culture by showing how the assumptions of sciences are shared with the general culture of the time period. An analysis of this sort may appear at first to be an intellectual endeavor of little import. However, as I shall argue, by taking the distant view of the history of scientific cultures, it is possible to develop a more critical stance toward today's theories of material and social worlds.

Of Paradigms and Epistemes

As I described the culture concept in the introduction, it is possible to compare cultures across time as well as space. The previous chapter took some steps in the direction of an analysis of early modern science in terms of its temporal culture. In this section I will elaborate the idea of temporal cultures in more detail.

In the book *The Order of Things* (*Les Mots et les choses*) the French historian Michel Foucault uses the term *episteme* (the Greek word for "knowledge") to describe the "epistemological grid" of a time period. The term will be elaborated in detail below; for now it can be approximated as the assumptions about knowledge, method, and theory that in any given time period are shared across "discursive formations" (which as a first approximation can be translated as "disciplines"). For example, nineteenth-century disciplines that focussed on life, language, and wealth (the forerunners of biology, linguistics, and economics) shared a way of seeing the world that relied on finding analogies among related elements in temporal narratives. That commonality is, as a first approximation, an example of an "epistemic" level of knowledge.

Foucault's episteme is quite a bit broader than the similar and much-discussed concept of the paradigm, which was popularized by the American philosopher and historian of science Thomas Kuhn. In *The Structure of Scientific Revolutions* Kuhn uses the term *paradigm* in several different ways, but it would be fair to say that followers of Kuhn often use the term to signal the underlying assumptions, categories, and frameworks of a single scientific discipline at any given period. Paradigms are general scientific frameworks that can undergo dramatic revolutions or shifts, as in the case of the transition from the geocentric to the heliocentric model of the solar system or the Newtonian to the Einsteinian views of space and time.

Although Kuhn and Foucault have very different views of the history of scientific knowledge, they both characterize their object in terms of fairly abrupt shifts or "ruptures." Thus, their position can be contrasted with other historians of science, such as Pierre Duhem and Lynn Thorndike, who have emphasized gradual change. Likewise, there is a history to the idea that history operates through relatively abrupt changes. In the history of science the idea has appeared in the work of thinkers such as Gaston Bachelard, and for history in general the idea of radical historical ruptures can be traced back at least to the concept of revolutionary change in the work of Karl Marx. There are fairly obvious ideological resonances with a theory of history that is characterized by ruptures versus one characterized by gradual change. The theory of ruptures argues for a world in which radical change is possible. Therefore, in a general sort of way such a way of thinking about history *may* be connected with progressive social programs. At the same time, however, I have been impressed with the frequently almost phallic descriptions of new forms of thought that are said to "emerge rapidly." The very idea of "ruptures" suggested a tearing through that might be associated with the masculine act of penetration, as in Bacon's description of science as the penetration of the secrets of a female nature. Of course, gradualist schemes are also able to emphasize "breakthroughs" on a smaller scale, and therefore they may also participate in the same sort of imagery.

In short, even the theories and metaphors used to interpret histories and theories are rooted in history and culture. In this case a theory of ruptures may have relatively more masculinist connotations, but it may also coincide with a view of history that resonates with some radical political programs. In choosing the alternative term *temporal cultures,* I will attempt to negotiate a middle ground between abrupt and gradual views of change. I see temporal cultures as interwoven layers that nonetheless are analytically distinguishable. However, before moving on to an outline of temporal cultures, it is helpful to spell out in some detail the achievements of Foucault's analysis, which will serve as a point of departure for the analysis of temporal cultures.

Foucault's Order of Things

Although I have begun with a move that points to a similarity between Kuhn and Foucault on the question of historical ruptures, there are significant differences between them. Whereas Kuhn considers a number of cases of paradigm shifts or revolutions, Foucault focuses on more general shifts at the beginning and end of what I call the early modern period. (Foucault calls it the "classical period," as in the classical music of Mozart.) Because Foucault's analysis is considerably broader than that of Kuhn, and in my opinion more sophisticated, I will work from epistemes as a starting point.

Foucault locates the classical episteme temporally after the Renaissance and Reformation, as discussed in the previous chapter, and before the French Revolution and the industrial revolution. Roughly, then, it covers the period from 1600 to 1800. In many of his other studies Foucault locates similar ruptures in social practices and technical discourses around the watershed point of the French and industrial revolutions of the late eighteenth and early nineteenth centuries. Thus, his studies cover the well-trodden terrain of early modern Western intellectual and social history. However, he leaves behind him a landscape that will never look quite the same again.

In *The Order of Things* Foucault's comparisons operate along two dimensions: across time periods and across scientific disciplines within a time period. He focuses on three disciplines—the sciences of language, wealth, and life. To understand better the classical period, he returns to the Renaissance episteme. He argues that the preclassical episteme emphasized an approach to knowledge based on symbols as part of the natural landscape such that everything was somehow secretly connected, as in the case of hermeticist thought. Thinkers saw parallels between the earth and the heavens, or they read secret divine meanings in inspired texts. Astrology, for example, operated on the basis of resemblances between the celestial and the terrestrial world.

In contrast, the classical episteme following the scientific revolution viewed symbols more as names; in other words, classical thinkers were more concerned with *representing* nature than with finding secret *resemblances*. Furthermore, the sciences of the classical episteme were organized around taxonomies that classified elements into a universal scheme according to relations of similarity and difference. The classical sciences tended to represent knowledge in terms of tables of items that could be measured and ordered (one sense of the "order of things" of the book's title). In some cases, such as Newtonian physics, the principles of ordering could be stated formally as mathematical laws. If, like most historians, one focuses on the scientific revolution of astronomy and Newtonian mechanics, then it is possible to describe its underlying principles in terms of mechanism and mathematical formalism. However, Foucault focuses instead on the sciences of language, life, and wealth, which during the seventeenth and eighteenth century did not undergo a similar transformation toward mechanism and mathematical formalism. Foucault therefore argues that the commonly held characterization of modern science as a combination of mechanistic and mathematical approaches to nature is a special case of a general approach to natural knowledge that was based on the principles of ordering and classification (1970:56).

Although Foucault's work shares with standard histories of science the concept of a radical and sudden change of thinking that occurred roughly at the beginning of the seventeenth century, he characterizes the nature of the change differently. He also argues that the sciences of the classical period should not be interpreted as modern in the sense of the sciences of the nineteenth or twentieth centuries. (In fact, Foucault and his followers would probably argue against the widespread characterization of that period as "early modern.") Foucault instead posits a second rupture at the end of the eighteenth century, where he locates the origins of a second episteme for which the assumptions still characterize, more or less, the sciences of today. The straightforward tables of representation that characterized the classical episteme became increasingly problematic as philosophers and others questioned the conditions of representation. As representation became problematic, the human sciences emerged to study "man" as knowing subject and object of knowledge. In general, tables of representation gave way to sciences that examined relations among the elements. Classification systems were made according to how structures function, particularly over time. Thus, the field of comparative languages or philology emerged, in which languages were classified in terms of comparative grammatical structures that were located in historically changing languages. In biology taxonomies were reorganized more along the lines of the functions of anatomical features and organ systems, and scientists began to examine those features and functions in terms of narratives of evolution. Finally, in economics value came to be seen as a function of labor, and the study of the historical dynamics of economies as systems of production became possible.

Temporal Cultures

Foucault's concept of epistemes is important in a number of ways. Unlike Kuhn's idea of paradigms, the episteme is transdisciplinary. Foucault also shifts the analysis of the history of science away from physics and the exact sciences, and he questions narratives of scientific history that locate origins in the scientific revolution of the seventeenth century. Still, Foucault's epistemic analysis is not the only way to think about transdisciplinary structures in the sciences. In this chapter I will develop a complementary form of analysis that connects structural changes in the content of science with those in society: an analysis of temporal cultures.

In *The Order of Things* Foucault carefully limited his analysis to three fields of inquiry, and in his other studies of medicine, psychiatry, theories of sexuality, criminology, and administration he is careful to delimit the scope of inquiry. Perhaps in conflict with Foucault and his followers, I assume that there are also intellectually rigorous and rewarding ways of examining transdisciplinary structures that are located at a cultural rather than an epistemic level. The analysis of temporal cultures moves across a wider range of disciplines and situates those transdisciplinary patterns/structures/networks as aspects of the broader culture. The interpretive project is difficult because it operates at a high level of generalization. That level of generalization makes it necessary to suspend, at least temporarily, the analysis of countercurrents, exceptions, and anomalies. It also requires a panoramic view of the general history that is difficult if not impossible to achieve in a world of fragmented expertise. Notwithstanding the risks, the project is worthwhile not only because it brings a broader cultural perspective to the problem but also because it can help frame a critical inspection and diagnosis of contemporary theories of society.

Before proceeding, it is necessary to begin with some caveats about what I will not assume in the analysis of temporal cultures. Those caveats may help avoid some possible misreadings of the analysis that follows. First, although the demarcation of a historical period punctuated by wars or other major events is a useful way in to the analysis of temporal cultures, the two projects are not the same. Temporal cultures are useful but fuzzy constructions that have amorphous temporal boundaries, can be subdivided, and vary temporally across disciplines and social domains. Furthermore, temporal cultures are better thought of as branching differentiations in the evolution of cultural forms than as successive historical layers. In other words, the older forms continue to exist and develop alongside the more recent ones. Thermodynamics, for example, may be in many ways a science of the nineteenth century, but research in the field continues in subsequent periods and becomes interwoven in subsequent developments.

Second, the analysis of cultural structures or forms is more than a thematic analysis of ideas of the type offered by Gerald Holton in *Thematic Origins of Scientific Thought.* Unlike the historian of ideas, I am suggesting an anthropological analysis of culture that locates general structures that scientific ideas share with social practices and discourses in a wide variety of cultural domains. Ideas do not float independently of social structure. At the same time, I also do not assume a base-superstructure kind of causality of the type commonly referred to as vulgar Marxism. In other words, ideas are not the dependent variable that is shaped by the independent variable of social structure, which in turn is shaped by technological "forces of production." Thus, the invention of the steam engine or internal combustion engine are not ultimately determining events that can be given a date that marks the beginning of a new period. Notwithstanding the romantic cult of the genius scientist and inventor, technical innovation is every bit as much a cultural process as other social actions. Technology, social structure, and ideas mutually influence each other and share common cultural structures. Although it seems clear from the historical record and daily news that in capitalist societies the capitalist class exerts hegemony over what gets said and produced, the form of that influence is rooted in a broad cultural order for which its participants are not likely to be entirely conscious. Even when the rich and powerful make history, they do not do it exactly as they please, because they make history in sociocultural structures that they may reproduce, willingly or unwillingly.

Third, although the analysis runs the risk of becoming a template that is imposed on historical evidence, it can be challenged empirically. However, to support or undermine the analysis, it is of no avail to point out that X concept or theory appeared too soon or too late in Y discipline, or that the transformations do not occur at the same time across disciplines. Nor does one challenge the analysis by pointing out counterthemes or counterstructures (which could be turned around to confirm, in a dialectical way, the general argument) or by noticing absences of the structures in some disciplines or cultural domains. Rather, one should point to alternative structures of temporal culture that could better fit the available empirical evidence. To make the analysis, I have continually rejected hypotheses of this or that cultural structure until I have arrived at one of sufficient generality to cover the material within my grasp. The analysis is therefore empirical, and it can be replaced by better analyses or complementary ones that propose other structures.

Fourth, although I am explicitly working at a general cultural level of analysis, I am not arguing that everything reduces to a single cultural pattern or structure. There are other general structures that crosscut many of the same cultural domains, although I think they lack the generality of the ones that I am proposing. For example, Foucault locates the break between the classical period and the nineteenth century in the appearance of "man" as both the knowing subject and as an object of scientific knowledge via the appearance of the social sciences. Likewise, the philosopher Jean-François Lyotard describes a general pattern in postmodern culture as "incredulity toward metanarratives," especially of the progressive type that were common in the nineteenth century. A third

TABLE 4.1
Temporal Cultures Among the Sciences

FIELD OF KNOWLEDGE	CLASSICAL *Classification, Structure*	PROGRESSIVE *Evolutionary, temporal dynamics*
Physics	Static world characterized by immutable laws (Newton, Boyle, etc.)	Cosmology (e.g., Laplace), thermodynamics, electromagnetism
Biology	Taxonomies, anatomy	Evolution, bacteriology, embryology
Economics	Analysis of wealth	Political economy
Linguistics/Language	General grammars	Philology
Psychology	Associational psychologies	Comparative/folk psychologies, romantic psychologies of the unconscious
Social Sciences	Social contract theory	Social Darwinism, cultural evolution

FIELD OF KNOWLEDGE	MODERNIST *Closed Systems, equilibrium*	POSTMODERN *Open networks, self-organization*
Physics	Early quantum mechanics	Gauge theories, nonlinear dynamics
Biology	Physiology of homeostasis, population equilibrium theory, genetics	Psychoneuroimmunology, molecular biology, nonlinear evolutionary models
Economics	Market and general equilibrium (Keynesian) theory	Increasing returns theory
Psychology	Dynamic psychiatries, learning theory	Consciousness studies, cognitive science
Linguistics/Language	Syntax and semantics, structuralism	Pragmatics, deconstruction
Social Sciences	Functionalism, culturalism, structuralism	Constructivism, reflexivity, poststructuralism

example is the work of the anthropologist Louis Dumont, who has traced the development of individualist values across a number of intellectual and social domains in modern Europe. At a more local level, discipline-specific ideas such as quantum theory can be situated in local cultural currents, as in the work of the historian Paul Forman on physics in Weimar Germany.

Fifth, the analysis of temporal cultures does not assume a single scheme that applies throughout the world. Rather, the analysis addresses patterns within specific societies at a point in their history. The analysis is also limited to the most developed regions of northwestern Europe, and secondarily to North America, other parts of Europe, and other areas of the world that are participating in the changes discussed. Outside the developing centers of the world system the periodization may break down, lag, or simply take other forms. Thus, the analysis of temporal cultures is a culturally specific formulation. It is also class specific—in other words, even in the most advanced societies not everyone lives in the same temporal cultures. As Octavio Paz has noted for Latin America, social strata in society often correspond to historical or temporal layers.

The Analysis of Temporal Cultures

The analysis of temporal cultures that I will sketch out here begins with Foucault's work but approaches the topic from a vantage point informed by an anthropological concept of culture. Some of the terminology will be different from that of the Foucauldians; for example, I use the term *modern* to refer to the period (for Western Europe) after roughly 1600 rather than after 1800. Within the modern period I distinguish four other temporal cultures. The *classical* or *early modern* period covers the scientific revolution through the Age of Enlightenment, a period that is framed by the Reformation and the British/Dutch revolutions on one side and the French/American/industrial revolutions on the other side. I cut the loaf of history into three other divisions: Progressivism (which spreads out on either side from the high point of the 1830s to 1850s), modernism (which spreads out from the 1920s), and postmodernism, in which we are presently muddling our way through (see table 4.1).

Progressivism

As an anthropologist thinking about history, I ask the question, "What types of societies were emerging during this period and how were they structured at a cultural level?" Thus, for some historians, I will show an annoying lack of concern with dates. The progressivist mood of the nineteenth century emerges in the late eighteenth century, with the American and French revolutions serving as one convenient marker of the transition, and it is still evident in some narratives of technology and society today. During the high tide of progressivist culture in the nineteenth century, utopian projects abound: people dream of building a new society through rationalist revolutionary politics, liberal social reforms, or radical communal organizations. Revolutionary and reform movements proliferate in what the sociologist Talcott Parsons in *The Evolution of Societies* has called the "Age of Revolutions." In some countries, such as the United States and Great Britain, political change takes the shape of reform movements, such as the anticorn law movement, Jacksonian democracy, and abolitionist movements. On the continent political movements are more revolutionary.

The "Age of Revolutions" is a convenient label because it can be extended to the economic and technological domain as well. During the Industrial Revolution capitalist production crystallizes into textile and iron mills, and steam engines are put to work in building railroads and industrial goods. The enclosure movement forces peasants and small landholders into the cities where they become wage laborers in factories. Just as the bourgeoisie emerged in the early modern period, so the working class emerges during the early parts of the nineteenth century. With the working class comes socialism, both utopian and revolutionary. Politics, like literature, is romantic: it looks to an ideal past to provide models for a more "rational" future.

Ecologically, the cultures of this period have an expansive orientation. For the New World countries, Russia, Australia, and some other lands, the relationship to nature is still an unclosed frontier. The European powers have not yet carved up the African continent and some other areas of the world where they are still in the process of exploration. Westerners envision the world as still having empty space and room for expansion— although of course the space is only empty from the perspective of the invading European populations.

Nineteenth-century science shares with the culture of revolutions and expansion a concern with history, progress, origins, genesis, development, and transformation. Although the term *progressive* is sometimes used in the United States for a later development in party politics, the term serves as an overall designator of the period because it flags the importance of time and progress as a cultural form. During this era history emerges as a sustained and theorized field of inquiry. Hegel, Comte, and, to some extent, Marx are examples of thinkers who theorize human history at a universal and general level, and other, lesser-known figures such as Ranke and Droysen theorize history as a field of inquiry. In other human sciences there is a similar concern with origins and development. For example, political economists postulate the temporal laws governing the accumulation of wealth or growth of population. Notwithstanding the great divergence between the liberal political economist John Stuart Mill and his radical contemporary Karl Marx, both construct the economic in terms of laws governing historical development. Likewise, notions of a primordial unconscious emerge in the psychologies of the period, succeeding the work of Mesmer and his followers. As among some of the romantic writers, there is a concern with a deeper side of the soul that carries with it hidden truth and powers. In language studies, folklore, and anthropology scholars search for the primordial and weave narratives of the evolution and history of languages and narrative forms.

For the natural world geology and cosmology emerge as sciences that theorize the origins and development of the earth and the universe. Just as those new sciences displace biblical creation stories, so bacteriology and the discovery of microorganisms replace the theory of spontaneous generation. Embryology enters a scientific phase based on detailed description. Physics also becomes concerned with genesis and origins. Thermodynamics replaces the reversible, clockwork world of Newton with one of irreversible entropy. Even the new experiments with electricity and magnetism show a concern with how one force can generate the other.

For the natural sciences, however, the key figure has always been Darwin, who is often held up as *the* nineteenth-century scientist. In the old order of the medieval world, both nature and society were organized according to a principle that the historian Arthur O. Lovejoy called a "great chain of being," a natural hierarchy or pyramid that ran from God at the top through the king and the various social classes to the various species in the plant and animal kingdoms. Buffon, Lamarck, Darwin, and other evolutionary thinkers opened up a different world in which relations among the species were fluid and subject to change. Upward mobility, reform, and revolution entered the world of life. Recall Marx's comment to Engels about how Darwin's nature mirrored capitalistic competition and struggles for existence.

What is perhaps less appreciated about the coconstruction of nature and society in the nineteenth century is what I call the "boomerang" of technototemism. Nineteenth-century thinkers built sciences that constructed nature through stories of origins and genesis: geology, cosmology, bacteriology, natural history, thermodynamics, and so on. Yet, because the cultural basis of those sciences is lost through an ideology of transcendental geniuses and methods, it is sometimes forgotten how much the scientists of the nineteenth century saw the world through the same lenses their contemporaries were wearing. Furthermore, it is only a small step before those lenses are turned back on society in the late nineteenth-century wave of social evolutionist theorizing by Spencer, Morgan, Tylor, Frazer, and others. Just as doctrines of Social Darwinism legitimated class privilege, so theories of cultural evolution legitimated colonial expansion. The mystification of the boomerang of techno-totemism provides a cautionary lesson for us today.

Modernism

When does the progressive period end? Perhaps it is better to think of it not as coming to an end but as differentiating such that modernist cultural forms gradually achieve more prominence. One might think of progressive temporal structures as a cultural network that expands at different rates through the various disciplines of science and domains of culture. The earlier structures do not disappear in subsequent temporal cultures; rather, they become submerged or merged into the new structures of emergent temporal cultures.

The various events of the 1860s and 1870s—the Austro-Prussian War, the Franco-Prussian War, the British Reform Bill of 1867, the Paris Commune, the American Civil War, and the liberation of slaves in the New World (which in some Latin American countries is delayed until nearly the end of the century)—mark the emergence of the modernist period. In *Darwin, Marx, Wagner* the intellectual historian Jacques Barzun characterizes the nineteenth century as a 125-year period that is epitomized by the three thinkers of his title in the year 1859 and ends only with the First World War. Likewise, in *The Condition of Postmodernity* the social theorist David Harvey locates a major intellectual break in the years immediately preceding the First World War. However, by the 1870s, and especially from the 1890s, artistic, literary, and technological innovations were beginning to show modernist structures.

In the technical and economic arenas there is a series of gradual displacements sometimes known as the second industrial revolution: iron production gives way to steel; railroads to roads and airways; telegraphs to telephones; and steam engines to power by electricity, internal combustion, and (by the mid-twentieth century) nuclear energy. It is an age of robber barons and monopoly capital, with militant trade unionism and nationalizations as the counterbalancing forces. The changes in the polity and economy may have partly driven changes in other sectors, but all domains are interacting in the making of an overall modernist cultural logic. Thus, the basic ecology and structure of Western societies, rather than merely the political and economic domain, is changing. Nature remains external to society as it had been in the nineteenth century, but nature is less and less the vast empty space of the frontier or the unexplored continent. As the frontiers close, Africa is carved up, the golden spike is driven into the American continental railroad, the electrical grids connect with each other, and the first transatlantic cables and flights are completed, a logic of ongoing expansion closes in on itself into a world of dynamic closed systems. As capitalism gives way to imperialism, the first signs of the global society emerge. With that new society comes the terrible downside of global warfare and destruction of the human project.

At a political level the modernist period is less an age of revolutions than an age of world wars. With the formation of the European states of Germany and Italy, conflict is exported—at least temporarily—to Africa. The last decades of the nineteenth century are the age of empires: large and unwieldy political systems subject to disequilibrium due to internal economic dislocations or to disruptions in the delicate international balance of power. To ensure domestic stability the regulatory agencies of the welfare state emerge, and to ensure international stability the international organization emerges. Although balance of power politics has a long history in Europe, it achieves a new importance because warfare is no longer local and is therefore increasingly unthinkable. From this perspective the cold war is a continuation of the same cultural-political logic. Under the Marshall Plan Germany and Japan are rebuilt as buffer states against the Soviet empire, and the old balance of power politics continues on a larger and more terrifying scale as the balance of nuclear terror.

As systems close in on themselves in ever greater cycles of complexity, the capitalists and political leaders of the period turn increasingly toward experts to equilibrate their worlds. A new class emerges—or, if one insists on the Marxist definition of class as a specific relationship to the mode of production—a new *stratum* emerges: professionals, managers, and other experts. The workplace becomes more rationalized through the advent of assembly line production and new management principles known as Taylorism. Unlike the mill, the assembly line factory exists as a delicately balanced closed system that shuts down when the balance between inputs and outputs is disturbed. As long as the parts arrive and everything is humming along, equilibrium is established and the assembly line continues. The production process does not change unless it is closed down temporarily because of exogenous factors. Viewed culturally, the assembly line—like the international system of empire states—is part of what could be called a second enclosure movement, to suggest a limited parallel with the enclosures of the industrial revolution.

Foucault mentions a series of parallel transitions in the last chapter of *The Order of Things*, where he discusses the emergence of sociology, psychology, and ethnology (anthropology). One sign of a new temporal culture is certainly the proliferation of new scientific disciplines, and the new human sciences of the modernist period are a great example. During that period, he argues, the ideas of "norm, rule, and system" came to inform the human sciences. The more general cultural analysis that I am suggesting would argue that those disciplinary-specific concepts can be generalized to a pattern of closed systems marked by equilibrium dynamics. Although the ideas of system and equilibrium are by no means new in the modernist period, the concern with synchronic, closed systems and dynamic equilibrium conditions is especially characteristic of modernist culture, both in the intellectual domain and in the technological, economic, and political domains.

I use the term *modernist* because it is now generally used in the humanities to refer to the arts, literature, and architecture that flourished in this period. In art modernism remains representational—of ideas, the world, or moods (as in some abstract art)—unlike the postmodern versions that the culture critic Frederic Jameson has described as representations of representations in a play of surfaces. In fiction modernism may be realistic or experimental, but in general it retains characters and plot in contrast to many of the experiments of postmodern fiction. Mysteries are solved and plots are closed, unlike postmodern fiction which tends to open up into ever widening circles of fantasy or paranoia. Perhaps the clearest image of modernist aesthetics is the architecture and design of cement and steel functionalism: the buildings of Le Corbusier, Niermeyer, or Wright, and the general vision of phallic skyscrapers, planned cities, streamlined transportation vehicles, and seamless highways.

In modernist science and thought nineteenth-century concerns of history, progress, evolution, and development are transformed into dynamic but relatively closed systems that operate according to a logic of equilibrium and homeostasis. Change occurs through the exogenous variables that percolate across systems, disturb structures, and are resolved in new equilibrium states. The concepts of equilibrium and homeostasis emerge most explicitly in the sciences of production and life, but I shall argue that similar ideas inform theory in a number of other disciplines as well. It is suggestive, too, that the disciplinary names are changing: political economy, for example, becomes economics. In that field Alfred Marshall develops a theory of dynamic (or long-term) equilibrium that alters the static models of nineteenth-century economic theory. His pupil John Maynard Keynes extends that framework with the theory that the economy can achieve at a macrolevel more than one point of equilibrium, and not necessarily at full employment. As a result government intervention becomes necessary to spur the economy to a different equilibrium point, and economic theory dovetails with a view of the state as having a regulatory role in the economy and society.

In biology equilibrium theory is expounded in both physiology and the syntheses of genetics and evolutionary theory. The physiologist Walter Cannon coins the term *homeostasis* to describe self-regulation in neurophysiological processes, and he argues for a New Deal-type regulation of the economy as a parallel to the regulation of the body by the autonomic nervous system. As in economics, the concept of equilibrium is dynamic: the organism is constantly changing to adapt to internal and external processes and to maintain a steady state. In evolutionary theory and genetics, the Hardy-Weinberg Equilibrium Principle posits a genetic equilibrium under conditions of random mating or crossbreeding. In other words, under those random conditions, the proportion of heterozygotes (impure or mixed forms) remains constant. In the synthesis of the 1930s evolutionary theory is still part of the intellectual landscape, but this science of the progressive temporal culture is reconstructed under the new equilibrium principles of population genetics.

Biology and economics were influential sources of metaphors for some of the modernist projects in psychology and psychiatry. In general, the emerging dynamic psychiatries articulated systems that involved relations between consciousness and the un/subconscious in contrast to the romantic views of primal unconsciousness that were more popular in the nineteenth century. Of course, aspects of the older models remained.

For example, Sigmund Freud's constructions of individual and collective pasts draw heavily on metaphors of archaeology, and Carl Jung's collective unconscious points back to nineteenth-century views. However, both psychiatrists also analyze the mind in terms of dynamic systems. For Freud the dream is a compromise formation between the need to sleep and the repressed materials that have been released due to the relaxation of the censor during sleep. Freud's later structural model (id-ego-superego) emphasizes a psyche that is composed of relations among various components, such as defense mechanisms, drives, and restrictions posed by the superego. The ego psychology of Freud's later life and many of his students, as well as many of the ideas of dynamic psychiatry in general, operates implicitly or explicitly with concepts of equilibrium. The ego is a mechanism that adjusts various opposing forces and moves toward an equilibrium, much as a market does. Symptoms are an expression of a system that is out of equilibrium or locked into a pathological equilibrium state because of trauma. Mental illness becomes "disequilibrium," and therapy is a means for restoring the psyche to a healthy state of equilibrium. Jung's theory of compensation provides another example of equilibrium concepts in modernist clinical psychology.

Other areas of psychology also build concepts of closed systems and equilibrium into their theories. For example, learning theory and operant conditioning assumes a dynamic of equilibrium conditions and disturbances. An organism is assumed to be in a state of equilibrium when a conditioned stimulus is given. At first the organism responds with an unconditioned response that suggests a system out of equilibrium. Over time the organism adjusts to the conditioned stimulus, and eventually it achieves a new equilibrium in the form of a conditioned response. As the Russian psychologist-physiologist Ivan Pavlov wrote, "Conditioned reflexes make possible the establishment of the most delicate and exact equilibrium between the complex organism and the surroundings" (1928:354).

Closed, synchronic systems characterized by equilibrium dynamics also appear in the exact sciences. Although I am inclined to interpret some theories as suggestions of what might today be recognized as postmodern—such as Gödel's incompleteness theorem and Heisenberg's uncertainty principle—some theories in physics show a characteristically modernist style of closed systems and equilibrium principles. It is worth recalling that many physicists during the last decades of the nineteenth century thought of their field as having exhausted itself. However, dramatic change came rapidly with the discovery of the electron and the emergence of a new layer of analysis. In the early decades of the twentieth century the new fields of atomic and particle physics emerge from cosmology, thermodynamics, electromagnetism, and other physical sciences that were elaborated and/or developed in the nineteenth century. At least some of the models of modernist atomic and subatomic physics are structurally similar to those of biology and economics. For example, the Rutherford-Bohr model posits atoms with shells that formed stable positions for the electrons. Emissions or absorptions of radiation correspond to changes in stationary states, in effect disruptions of equilibrium conditions. Likewise, in chemistry a discourse of equilibrium emerges in the 1920s through theories of reversible chemical reaction and equilibrium points in condensation and vaporization.

Discourses and theories based on closed systems with equilibrium dynamics are, therefore, not equally prevalent in all disciplines. In language studies, for example, there is much less evidence for the importance of equilibrium dynamics. However, the emergence of linguistics from philology is accompanied by a reconfiguration of the study of languages as closed, structured systems rather than historically evolving forms. On this point one of the founders of modern linguistics, Ferdinand de Saussure, distinguishes between a diachronic, or cross-time, perspective and a synchronic perspective that focused on the structures of a language. His synchronic approach became highly influential in modernist anthropology and literary studies, but it is in his lesser-known diachronic approach that something approximating the concept of equilibrium can be found. At the end of his famous *General Course in Linguistics* Saussure argues that a principle of "intercourse"—in a sense, a desire and need to communicate—is what spreads language and gives it unity. The principle toward homogenization counteracts the varying centrifugal forces that tend to pulverize a language into dialects. The grammar of a language at any given time, then, is in effect an equilibrium state that reflects the balances of forces toward change and uniformity that operate on a language.

What about modernist social theory? Early modernist theorists such as Boas, Durkheim, and Weber were still very immersed in the evolutionary questions of the nineteenth century. At the same time they opened the way for more synchronic and systems-oriented approaches to social analysis, which are characteristic of modernist thought. In the process the boomerang of technototemism appears again, for the modernist theorists—and especially their students—borrow heavily from linguistics, economics, and biology. Social Darwinism is displaced, but the new social theories suffer from other distortions in their construction of the social world.

One important transition figure is Franz Boas, a German immigrant to the United States who is sometimes called the founder of American anthropology. Deeply immersed in the project of writing the histories of nonliterate societies and cultures, he developed an analysis of discrete culture "traits" that could pass from one culture to another like genetic traits in a biological population. His project of historical reconstruction through the analysis of culture traits resonated with the evolutionary mood of the nineteenth century. However, Boas's project was largely defined in reaction to nineteenth-century anthropologists, Social Darwinists, and others, who posited universal sequences of evolutionary stages through which all societies would pass. In place of the evolutionary framework Boas, and especially his students (such as Alfred Kroeber and Ruth Benedict), developed the modern culture concept, which examined cultures as integrated wholes with distinct overall patterns.

Like Boas, the sociologists Max Weber and Emile Durkheim had one foot in the nineteenth century through their concern with general theories of historical development. As discussed in the previous chapter, Weber wrote a great deal about the general historical processes of rationalization and bureaucratization. However, Weber also opened up ways of looking at society that posited mechanisms of feedback in synchronic systems. For example, in the opening chapters of *Economy and Society* he argues that a political system can change through the event of charismatic leadership, but usually upon the death of a charismatic leader there followed a period in which the charisma was routinized. The "routinization of charisma" therefore posits a cycle of disturbance and stabilization akin to biological homeostasis and economic equilibrium.

Durkheim's discussions of religion, suicide, and social solidarity also provide an analysis of the social as a systematic domain of its own that cannot be reduced to the individual, much as Boas and his students do with the culture concept. In *The Divison of Labor in Society* Durkheim applies biological metaphors to analyze society in terms of "function" and "organic solidarity." He poses the problem of society as one of achieving social integration (a form of equilibrium), and he argues that modern society is held together through its division of labor into discrete functions, much like organs in an organic system. He also outlines states of disequilibrium or social pathology. The picture emerges of society as a kind of organic being (Boas's student Alfred Kroeber actually used the word *superorganic*) that has a physiology and is subject to states of disequilibrium and pathology.

However, Durkheim and his school are complicated because their work also brings economic and linguistic metaphors into the study of society. For example, their analyses of exchange (as in Mauss's *The Gift*) and classification (as in Durkheim and Mauss's *Primitive Classification*) examine closed, synchronic systems with less reliance on metaphors of physiology and pathology. Both the Mauss essay and the Durkheim and Mauss project influenced anthropology and the late modernist work of Claude Lévi-Strauss, whose studies draw on exchange theory for the analysis of kinship and classification/semiotic theory for the analysis of myth. Although the concept of equilibrium is not a salient concern in Lévi-Strauss's structuralist anthropology, the modernist mood of closed systems and the tendency toward formal analysis characterizes much of his structuralist project. (At the same time his project of New World mythological analysis remains curiously unclosed and therefore suggests a transition toward postmodern perspectives on culture.)

In some of the social theorists of the mid-twentieth century, thinking in terms of equilibrium is much more explicit. For example, when American cultural anthropologists of the modernist period examined acculturation and culture contact, they viewed cultures as discrete systems—not as connected parts of a world system, transnational hybrids, or disjointed relationships between homeland and diaspora, as they are viewed among today's postmodern anthropologists. In the modernist models of acculturation culture change occurs through disruptions caused by innovation or culture contact. The response to culture contact is either to disintegrate or to achieve a new equilibrium state of integration. In a similar way, British social anthropologists examined rituals, feuds, and other social phenomena as self-correcting mechanisms that allowed social groups to blow off steam and express conflict without destabilizing the social order.

Modernist social thought receives perhaps its highest expression in the theories of the sociologist Talcott Parsons. Parsons drew on discussions of equilibrium in the work of the economist and social theorist Vilfredo Pareto and the biologist Lawrence Henderson, who like Cannon applied biological equilibrium theories to society. Parsons also drew on the mathematical models of cybernetics that posited systems with feedback that could be used to model equilibrium dynamics. He developed a complex and complete theory of society as a self-correcting system based on equilibrium principles. Consider the near carnivalization of the equilibrium concept in one statement by Parsons and his colleagues:

The social system's own equilibrium is itself made up of many subequilibriums within and cutting across one another, with numerous personality systems more or less in internal equilibrium, making up different equilibrated systems such as kinship groups, social strata, churches, sects, economic enterprises, and governmental bodies. All enter into a huge moving equilibrium in which instabilities in one subsystem in the personality or social sphere are communicated simultaneously to both levels, either disequilibrating the larger system, or part of it, until either a reequilibration takes place or the total equilibrium changes its form. (Parsons and Shils 1951:226–27)

The functionalism of Parsons, the structuralism of Lévi-Strauss, and the culturalism of the American anthropologists have all been subjected to heavy criticism by subsequent generations of social scientists. Just as the universal evolutionary sequences of the social evolutionists have been criticized for a conservative political bias, the static and ahistorical frameworks of structuralism, functionalism, and culturalism have been criticized for ruling out the disruptive role of actors who are able to become conscious of structures and remake their history. Although to contemporaries those theories may have seemed convincing—partly because they were so deeply enmeshed in modernist culture—the theories now seem to be very much products of their cultures (which, in a dialectical manner, those theories also helped to reproduce). The boomerang of technototemism, then, is not restricted to the early modern and progressive social theories. It continues in modernist science. The question emerges, does it also operate in today's "postmodern" theories of society?

Postmodernism

The present is always the most difficult epoch to interpret historically because it is so hard to know what will become a blip on the historical register and what will become an "event." Still, there is a growing consensus in social and cultural studies that we are in a new era. It remains a capitalist era, and in this sense we are still in a period that can be called modern and is connected, however remotely, with the seventeenth century. The idea of yet another bifurcation toward structures of postmodernity refers to changes in the world in the wake of World War II (as the historian Donna Haraway argues in *Simians, Cyborgs, and Women*) that gain increasing momentum after 1968, the Vietnam War, and the beginning of superpower détente (as the political economist David Harvey argues in *The Condition of Postmodernity*). By the 1990s the events of the information superhighway, the genome project, global warming, and the end of the cold war were markers of a world that appeared increasingly different from that of the modernist period.

Ecologically, the nature/society relationship no longer has the robustness of the two bounded systems that characterized previous temporal cultures. In other words, nature can no longer be conceptualized as an externality to the human system under analysis (as, for example, pollution is modeled in conventional neoclassical economics). Nature is no longer the great outdoors, because there is too much indoors out there. The boundary between human society and nature has become increasingly muddied as the public becomes aware of pollution, species extinction, global warming, overpopulation, and environmental catastrophe. It is not necessary to accept the idea of Gaia as a living bioenvironmental organism to recognize that there is still a great popular awareness of the interwoven aspects of human population, technology, the climate, and ecology. As the evening news portrays the worldwide death of frogs— metaphorical canaries in the global mine—popular consciousness takes a hard swallow and wonders how long the planet and with it the human project will survive. The HMS *Beagle* gives way to spaceship earth.

The breakdown of the nature/culture boundary is only one example of boundary transgressions that have come to replace more robustly demarcated systems. Politically, the postmodern world is characterized by the breakdown of the empire-state and nation-state. Although the British Commonwealth emerged during the modernist period, during the decades following World War II the dissolution of various empires— American, British, French, Portuguese, and finally Soviet—has been prominent in international politics. In the center countries the breakdown of the nation-state has continued through the emergence of trading blocks, multinational capital, transnational flows, and international organizations. In some postcolonial societies colonialism and neocolonialism have given way to large-scale industrialization, national bourgeoisies, and competition in the global economy. In other postcolonial societies the fragile nation-states that were left to replace the colonies of the empire-states have not been able to cope with postcolonial economic devastation and ethnic conflict, and there has been a complete breakdown of the nation-state. In the poor countries ethnic conflict, local fascism, and starvation-level poverty have led to huge transnational population flows and immigrant diasporas. In the rich countries population dislocations are mediated more through a rapidly changing global economy that results in unpredictable mass layoffs, runaway factories, or mean-and-lean downsizing. The anthropologist Arjun Appadurai (1990) characterizes the contemporary period as involving transnational flows of many types: people, capital, media, and ideologies. This rapidly changing world of multiple global flows seems increasingly chaotic, and nation-states and local communities struggle to maintain self-organization.

In the economic domain monopoly capital is giving way increasingly to what the economic historian Robert Reich in *The Work of Nations* has called "enterprise webs" and "global webs." In other words, old boundaries among the state, capital, labor, and the university are disturbed in an era of mergers, acquisitions, joint ventures, franchises, spin-offs, and partnerships driven by competition on a global scale. Likewise, as David Harvey has noted, financial markets are tuned to each other in global loops of self-reference. The sun may have set on the British empire of yesterday, but it never sets today on the global stock market. As Harvey also describes, the old order of assembly-line production and Taylorist managerial technologies has given way to flexible accumulation and new managerial technologies. Factories are able to change products quickly in order to adapt to rapidly changing markers, and workers are trained to know more than one job and to interact with management in the design of the production process. Thus, the boundary between production and marketing, inside and outside, is transgressed as firms set up internal structures that allow for rapid and flexible responses to changing external markets. Part of that flexibility includes transforming workers from "cogs" in the assembly line to "holons" of the production process. In other words, new managerial techniques court worker participation in decision making, and they train workers to have a more general knowledge of the production process so that they can move from one job to another as products change (Martin 1994).

In an era of global and flexible production the computer is everywhere: in transportation, communication, production, distribution, and consumption. The computer is *the* technology of an era of self-organization and boundary transgression; it is the steam engine or electric motor of the postmodernist temporal culture. It is also a highly flexible technology, capable of modernist and postmodern forms. Under the modernist guise the computer can operate as an assembly-line machine, as in the case of workers who enter data mechanically in assembly-line offices of row after row of clerical labor. Under the postmodern guise the computer is the machine of artificial intelligence and the site of virtual communities. As the portal to cyberspace, the computer is postmodern in that its virtual worlds transgress local boundaries and self-organize into constantly evolving virtual communities.

Just as the boundaries surrounding the production process are dissolving, so are those around the family and gender. In the new era women of the middle classes have entered the workforce in huge numbers. Services once performed within the domestic division of labor—cooking, cleaning, child care, home repair, lawn care—are being professionalized as the boundary between the domestic and outside worlds dissolves. At the same time the nuclear family and conventional marriage are giving way to flexible or alternative living configurations such as partnership, communal living, gay/lesbian couples, and single-parent households. In the highly mobile society of the United States this pattern is perhaps the most developed, for here it is not uncommon for individuals to move (literally) through life and recreate a virtual family (of friends and whatever relatives may be nearby) in each city where they live. Family and neighborhood ties are weakened and anonymity increases, but at the same time alternative living arrangements that were once taboo become more acceptable.

The changing social structure also includes the emergence of new classes or strata. In *The Work of Nations* Robert Reich has argued that a new type of entrepreneur is appearing, one who can mediate the interstices of the global webs. The new "class" of technical mediators or translators blends elements of the small shopkeeper/professional, the independent entrepreneur, and the manager/contract laborer. Unlike the old shopkeepers that Marx called the petty bourgeoisie, and unlike the updated version of shopkeepers sometimes called the professional class (such as doctors and lawyers), this new class is less rooted in a local community or locally based practice. Unlike managers, members of the new class of technical mediators are not bound to a specific firm, nor are they as interchangeable and replaceable. However, like the small

shopkeepers and entrepreneurs, the members of this new class are independent agents, and, like managers and professionals, they own and sell
expertise. Nevertheless, the expertise of the new class is different from
that of managers and professionals because the technical mediators
trade on national or even international individual name recognition.
Their name, not their educational degree, is their capital.

There is a diverse number of formulations of postmodernism, but
anthropological perspectives have not yet been thoroughly elaborated. I
suggest that at a general level one of the major patterns of postmodern
culture—from ecology and economics to art and science—is a tendency
away from structures involving finite systems governed by equilibrium
dynamics toward open systems (or "networks") characterized by boundary transgressions and governed more by a logic of self-organization or
reflexivity. Again, it is not so much a case that modernist systems and
dynamics are disappearing than it is a case of postmodern ones emerging.

In the natural sciences the postmodern moment is seen in the proliferation of theories of chaos, complexity, and second-order cybernetics.
In the famous fractals of chaos theory, iterative nonlinear equations generate self-organization through structures that replicate in endless levels
of similar difference. The underlying mathematics is simple in that it is
iterative: the result of the equation is reentered into the equation, and
the result of that equation is reentered into the same equation, and so
on. The resulting pattern seems to model very well a number of complex
natural phenomena, including heartbeat irrhythmia, weather patterns,
and state transitions of solid to liquid. When the strange attractors of
nonlinear systems (a kind of nonequilibrium equilibrium) are connected
with the modernist biology of the evolutionary synthesis and the programming capabilities of the computer, it becomes possible to model
ecosystem evolution, business cycles, and perhaps even human learning. Chaos and complexity theory—or better, nonlinear programming
models—are sweeping across the intellectual landscape in ways similar
to the equilibrium models of the modernist period.

However, postmodern theorizing is not limited to nonlinear modeling.
In biology the reflexive or self-referential tendency is most apparent in
new knowledges that are turning nature into an increasingly manufactured or cultural entity and evolution into a self-referential process. Molecular biology transforms the material world of genes into an informa-
tional world of transcriptions, translations, and computational processes. At the same time genetic engineering transforms the evolutionary

synthesis of the modernist period into a self-referential process. The results of the genome project—the scientific project to map human genes—make almost inevitable the process of continued and direct human interference in the evolutionary process. It is true that in the past human societies have interfered with evolutionary and genetic process-es through selective animal and plant breeding. Likewise, in the earlier part of the twentieth century the eugenics movement called for the improvement of society through selective breeding. However, as the anthropologist Paul Rabinow (1992a) has argued, in the eugenics move-ment nature was still outside society. In the new age of the genome pro-ject nature becomes a manufactured object that is no longer outside society. The division between nature and culture is much more prob-lematic with the advent of what Rabinow calls "biosociality." Perhaps nowhere is the new pattern more clear, as Rabinow notes, than in the case of genetic engineering of plants and animals. Genetic engineering also holds out the dangerous temptation of attempting a technological fix for the world's grave environmental crisis. For example, rather than stop acid rain, some people are stocking lakes with bioengineered fish that can survive better in the new toxic environment.

Related issues of reflexivity and the manufacture of nature have also emerged in late twentieth-century physics. Physicists now speak of a "November Revolution," which occurred in 1974 and was accompanied by major theoretical changes. By the end of the 1970s a new physics of quarks and gauge theory had come to replace the physics of the 1960s and previous decades. As the STS researcher Andrew Pickering argues, physicists have changed their experimental practice to one that recog-nizes more the constructed nature of their enterprise. "Experimentally," Pickering notes, "the new physics was 'theory-oriented': experimenters had come to eschew the common-sense approach of investigating the most conspicuous phenomena, and research traditions focused instead upon certain very rare processes" (1984:15). The transition toward a theory-oriented form of experimentation suggests that even in physics scientists are seeing nature as somewhat manufactured or shaped by theories rather than seeing their theories as merely being driven by experimental results. Furthermore, there is no longer any sense of clo-sure or of finding any ultimately foundational particles; the sense today is that the search for more infinitesimal levels of nature is only limited by the resolution powers of the available machines. Theoretically, nature itself begins to look like the fractals that replicate endlessly at each level of resolution. In the old Indian expression, the world rests on a turtle, which rests on another turtle, and after that, it is turtles all the way down.

Just as physicists have become more aware of the theory-driven nature of their enterprise and structures of self-replication, so have researchers in psychology, cognitive science, and artificial intelligence. New models of artificial intelligence borrow on nonlinear systems to break with modernist models of programming that involved a logic of a master program and subroutines. Virtual reality programs that put the users inside the computer create the possibility for a constructed, artificial world of the human-computer interface. Virtual reality and other explorations of cyberspace challenge the self/other distinction by making the world the product of the self's manipulations. In a similar way new forms of clinical psychology have shifted the focus away from the self as a closed system to an open network. Psychologies of consciousness have emerged as a site of ferment comparable to psychologies of the unconscious at the beginning of this century. For example, lucid dream research revolutionizes the dream from a "compromise formation" or equilibrium state that emerged through the play of forces in the dream work to a self-consciously generated altered state of consciousness. In this situation dream symbolism becomes symbolism about the self or even about the process itself of dreaming and dream control.

In medicine the human being is now accorded an increasing role in the transformation of the biomedical model into the biopsychosocial model of illness and healing. New disciplines such as psychoneuroimmunology situate biological processes (still governed by equilibrium conditions) in a complex network that opens out onto conscious intervention. As the anthropologist Emily Martin has shown, popular understandings of disease are increasingly colored by the discourse of immunology, which constructs a flexible and adaptable body like that of the new workplaces of flexible production (1992, 1994). Health is not an equilibrium point reached after the mechanisms of resistance overcome the disequilibrating forces of infection; rather, health is an emergent property of a complex set of interacting systems that includes, reflexively, state of mind. Mind, too, is an emergent property of neural networks, and it can be changed at any number of levels. Thus, it is now possible to shortcut the closed system of psychotherapy through drug-induced personality transformations that are claimed to accompany drugs such as Prozac.

In the social and human sciences developments from anthropology to philosophy have increasingly undermined the modernist idea of a human self that is somehow outside the system that it observes. Instead, analyses have tended to transgress the boundary between the observing self and the observed system. As this happens, the study of closed systems opens up to reflexive analyses that include the observer, those who observe the observer, and so on. Thus, open systems and reflexivity are interwoven features of the postmodern epistemic mood.

In my own corner of the social sciences the "new ethnographers" have focused increasing attention on their relationships with their informants and on the position of anthropology in historical structures of colonialism. Cultures are no longer isolated units that can be studied as discrete systems. Rather, they are part of an interlocking global network of local and transnational cultures that includes the culture of the observing anthropologist. The observed/observer relationship has led to renewed interest in a genre of ethnography known as the fieldwork account, in which anthropologists discuss in a more personal way their own position in the construction of the other culture. Likewise, in STS some researchers have focused on the reflexive issue of how social scientists' accounts of science are themselves socially constructed. Women anthropologists and feminist STS theoreticians have played an important role in this new, reflexive ethnography, although they have not always been given their fair share of credit. A related area of reflexive thinking is contemporary linguistics, where some studies in pragmatics (speech acts in context), intercultural communication, and sociolinguistics have provided the tools for a form of analysis that operates across linguistic barriers rather than within a systematic language. From anthropology, women's studies, and language studies have come carefully positioned analyses in which researchers are explicit about their own relationship to the material they study. The observing self is therefore analyzed in terms of the same matrices of race, class, and gender that are used to analyze the texts, practices, and technologies of the observed other.

The emergence of self-referential, recursive, positioned, and reflexive frameworks has swept across the various disciplines of the humanities and social sciences through a number of related movements: deconstruction, poststructuralism, feminist positionality theory, new ethnography, reflexive social studies of knowledge, intercultural communication studies, and so on. One marker of the emergence of something "new" on the intellectual landscape is that new disciplines (or, in keeping with the boundary-transgressing pattern, new "interdisciplines") are emerging and, to some extent, displacing the older disciplinary formations of the social sciences and humanities. During the modernist period linguistics, cultural anthropology (or ethnology), dynamic psychiatry, and economics emerged from philology, evolutionary anthropology, neurology, and political economy. Today a similar invention of new interdisciplines is emerging from literature, history, anthropology, sociology, linguistics, and psychology: cultural studies, various ethnic studies, women's studies, gay and lesbian studies, science and technology studies, cognitive science, pragmatics, intercultural communication studies, and so on. The emergence of new disciplines is one marker of a shift of temporal cultures. At the same time the interdisciplinary spirit of the new disciplines suggests the ways in which boundaries are transgressed more easily in a world of open systems.

Conclusion: A Cautionary Tale

I will end, in the literary tradition of the seventeenth-century American jeremiad, with a cautionary tale about postmodernism that may make more sense after having gone through the exercise of examining progressivism and modernism. Journalists now write breathless, best-selling books about the frontiers of contemporary science: molecular biology, artificial intelligence, chaos and complexity theory, grand unification theory, and so on. Patterns emerge across disciplines, and talk of transcendental, supracultural genius abounds. Yet, I would urge suspicion of the ways in which patterns get transformed into a quasitheological discourse about the acultural nature of new scientific truths. When I look at the beautiful, four-color pictures of computer-generated Mandelbrot sets, and when I read narratives of geniuses having epiphanies under the desert sun of Santa Fe, I do not see science finally unlocking elusive secrets of the "old one." I see white men constructing the technoscientific culture of postmodern global capitalism. To interpret their culture as transcendental is to take the first step toward mystification and the boomerang of technototemism.

In the New Age movement the boomerang effect is already well under way. Here, Lyotard's description of the postmodern condition as "incredulity toward metanarratives" is displaced by a credulity toward metascientific narratives that in some cases borders on the irrational. For example, Emily Martin warns of a new Social Darwinism that draws on immunology to legitimate narratives of who wins and loses in society (1994). She shows how New Age gurus provide methods for training the immune system to ensure survival in the era of AIDS and other immunological disorders. Likewise, in *Science in the New Age* I have shown how New Age intellectuals transmute physical theories of dissipative structures and nonlocality into claims of a new social world in which the part is a mirror for the whole. In this new world of the holographic universe, the science of emergent system properties sometimes gives way to a style of thought that is reminiscent of the old Renaissance episteme of resemblances. From the perspective of some of the New Age intellectuals, the new culture is not only postmodernist but postmodern in the deep sense of breaking with the fundamental assumptions of thought since the seventeenth century. An antirationalist undercurrent can be found in some of the formulations. In others, such as Marilyn Ferguson's *Aquarian Conspiracy*, discussions of the new holism as played out in a spirit of cooperation suggest a naive political analysis of the vagaries of postmodern capitalism. Such stories need to be inspected critically; boundary-transgression and the emergent properties of holographs and nonlinear systems do not imply that we are on the verge of living in a better society in which competitive hierarchies give way to cooperation and analytic, top-down thinking and social organization give way to synthetic, bottom-up alternatives. If anything, we are in an age of leaner, meaner, downsizing capitalism that appropriates a rhetoric of participation from labor in the interest of making higher profits.

Just as dangerous are more erudite versions of postmodern social theory that focus entirely on the reflexive question of how human scientists produce their own knowledge. Although critical self-inspection is necessary and healthy, some of the more extreme versions of deconstructive analysis can lead their practitioners away from social criticism to the whirlpool of self-reflection. Reflexivity and deconstruction sometimes get lost in solipsistic word games and epistemological paradoxes rather than considering relations between self and other at the level of communities or societies that are engaged in their own macrostructural power struggles. When those projects connect with a discourse of chaos or complexity theory, they may become even more seductive. I fear that a new wave of social chaos theory is about to sweep the popular consciousness, much as social evolutionism did in the nineteenth century or regulatory functionalism did in the mid-twentieth century. Just as Emily Martin warns of a new Social Darwinism in the popular applications of immunology, I would also caution against the new Social Darwinist undercurrents of some of the recent combinations of the evolutionary synthesis of the 1930s with nonlinear systems theory that are being applied to economic phenomena.

Finally, before claiming uncritically that "we" are living in a postmodern age, it is worth remembering that not everyone is included in that *we*. Cyberspace is an elite space, a playground for the privileged, as are the New Age weekend getaways for higher consciousness or immune-system training. There *is* a global glass ceiling, and for many in the world a large part of postmodern technoculture lies well above it. Molecular biology, the new physics, and nonlinear computer models are sites for the reproduction of big medicine, big science, and big technology. Yet, because many of the cutting edge fields are new and unstabilized, they are also sites where the decisions have not all been made and black boxes not firmly sealed shut.

A cultural and historical perspective such as the one I have outlined in this chapter can help make it possible to understand better the nature of the changes that are taking place. It also may help identify historical sites that are still in a relatively unstabilized state, as well as the structures and networks of an emerging (if not emergent) cultural order. At the same time a cultural perspective can help scientists, activists, citizens, and others to think through how "we" reproduce the emergent culture even as we seek to question and transform it.

5 | The Social Relations and Structures of Scientific and Technical Communities

Just as culture and politics play themselves out in the content of science and technology, they are also woven into their social relations and social structures. Most scientific disciplines and communities today are transnational in nature. There are physicists in Japan, Europe, India, Latin America, and so on, and they have transnational organizations, conferences, and electronic networks that sustain the international ties. One way of thinking about the transnational nature of scientific communities is to argue that they are supranational in the sense that national cultural differences are relatively unimportant in comparison with the shared features of the communities. That viewpoint is consistent with the general ideology that science and technology are supracultural and therefore somehow outside society and culture.

Although there are undoubtedly many supranational features to scientific communities, I will focus in this chapter on the less obvious ways in which they vary across cultures. To begin the discussion it might be fruitful to think of transnational scientific and technical communities as having some features in common with diasporas. In the classic case of the Jewish and African diasporas, an originary homeland serves as a point of reference for unique communities in Europe and the New World that are mixtures of the transnational diaspora culture and the local national culture in which they are located. To some extent international scientific and technical communities have an original homeland, usually Western Europe or North America, but the local communities also have features in common with the national cultures in which they are embedded. Likewise, many of the members of the international communities have spent a good deal of time—as graduate students, postdocs, or visiting researchers—in a few of the main international research centers in their field.

I have already made the argument that in several cases where social scientists and historians have asked the question, they have found that the content of science and technology differs across national boundaries. In this chapter I will continue the critique of supracultural views of science and technology by focusing on the question of how cultural differences work in the social relations and social structures of transnational scientific and technical communities. To do so I draw on two major literatures: research on intercultural communication (including gender and communication studies) and classical studies of national cultural differences in comparative sociology and cultural anthropology.

Conceptual Background

The comparative study of national cultural differences in the social sciences has a long and rich history. A frequent starting point is the work of the French historian and social observer Alexis de Tocqueville. In the middle of the nineteenth century de Tocqueville traveled to the United States and wrote *Democracy in America*, a marvelous comparative study of what today would be called the cultures of France and the United States. In many ways de Tocqueville was doing cultural anthropology before its time. De Tocqueville was so astute that many of his observations still ring true today, more than 150 years after he traveled through the United States. He attempted to go beyond the surface differences between French and American culture: the French do X *this* way, whereas Americans do it *that* way. Instead, he tried to describe what anthropologists would later call the themes, patterns, values, or structures of the two cultures. In the process he came upon some of the more salient values and structures of American culture, such as its almost obsessive individualism (in contrast to the importance of status groups, classes, and extended families in other societies). He also articulated Americans' concern with equality, including its contradictions (e.g., slavery) and its contrasts with the social pattern of hierarchy of European countries, with their aristocratic legacies. Although the United States has changed dramatically since de Tocqueville's time, it has also remained individualistic, at least in comparison with most other societies, and egalitarian, although often more in ideal than in practice.

In the twentieth century social scientists developed de Tocqueville's project into the field of comparative sociology and the cultural anthropology of national societies. The American sociologist Talcott Parsons, for example, developed a useful terminology that contrasts the "universalism" of modern societies with the "particularism" of traditional societies. Societies that tend to have a single value system across social situations are said to have "universalistic" values. In contrast, societies with different values for different social situations are said to be "particularistic." In a similar way the French anthropologist Louis Dumont explored and compared modern value systems based on equality and individualism with traditional ones based on hierarchy and what he called "holism," which contrasts with individualism in that the self concept is integrally linked to group membership.

Closely related to the project of the comparative study of social structures is the field of intercultural communication. The anthropologist Edward T. Hall developed a set of terms to characterize how cultures differ in their communication styles. Styles in which information is conveyed primarily in the verbal message are designated *low context,* whereas those in which a great deal of information is embedded in the context (the appearance, setting, and social relationship) are called *high context.* Cultures can then be arranged on a continuum from the low context to the high: German-Swiss, German, Scandinavian, American, French, English, Italian, Spanish, Greek, Arab, and Japanese. Cross-cutting this dimension is the orientation toward time. Polychronic cultures are people-oriented jugglers in which time depends on social relations, whereas monochronic cultures are schedule-oriented plodders who tend to do one thing at a time and view time as something that can be spent and wasted. Polychronic cultures tend to be high context and monochronic cultures tend to be low context, but in some cases (such as Japanese business settings) the high context culture is also monochronic.

I have only given here a rudimentary vocabulary for the cultural analysis of comparative social relations and social structures. The terminology provides a good starting point, but individual case studies will require more complicated terminology and distinctions. Thus, the terms should be taken as useful "ideal types," to borrow Max Weber's phrase. In other words, the terms are useful guideposts for understanding aspects of the social world, but the concrete cases never correspond exactly to the ideal types. In this chapter I shall introduce some specific cases. My goal is not to provide a comprehensive survey of all the different cultural styles in the social relations and structures of technical communities, but instead to suggest some basic concepts that are applicable to a number of other cases. The first section of the chapter deals with social relations, which will be limited here to issues of intercultural communication and their potential application to communication among scientists and engineers. The second section treats the issue of the comparative social structure of scientific communities, which will be limited to the social organization of a few academic research organizations.

Before beginning, some warnings are in order. First, because most of the readers of this volume will be Americans, and for many others (including myself) the United States is a relatively familiar culture (either through television or visits), I use American culture as a point of comparison. Like languages, national cultures can be thought of as belonging to families. Much of what can be said about American culture can be extended to other, related Anglophone societies such as Australia, Canada, New Zealand, Ireland, and Great Britain. It can also be extended to some of the Protestant cultures of Northern Europe. However, there are also significant differences, even between such closely related cultures as North European-descent populations in Canada and the United States. Thus, generalizations should be checked with observation and not made automatically.

A second warning is that I will use expressions such as "the Americans" or "the French" only as generalizations, as rough approximations that could later be refined (and sometimes overturned) as the focus narrows to specific social classes, genders, ethnic groups, and regions within each of the national cultures. The generalizations are based on long-term observations, and anyone who bothers to learn a foreign language and live in another country is likely to see how true they are. The use of such general categories by no means negates the enormous internal differences that can undermine any generalization. In short, the comparative study of national differences should be seen as a loose type of analysis that produces a sense of general tendencies, not iron laws of behavior.

The study of national cultural differences is therefore a dangerous business. It can quickly slip into confirmations of popular stereotypes, and stereotypes can quickly slide into racism. To apply general descriptions of cultures to individuals constitutes the "ecological fallacy," the fallacy of thinking that what applies to the group necessarily applies to the individual. It is better to think of the generalizations as characterizations of central tendencies in populations, with widespread variation around the central tendencies. The danger of the ecological fallacy can also be mitigated by anchoring generalizations in detailed observations based on social science methods such as ethnography, interviewing, and surveys. Used wisely, social science methods apply ideal types to promote cultural understanding, not stereotypes to promote ethnocentrism. Furthermore, there is a second type of danger that occurs if the analysis is completely rejected. *Not* to have a solid understanding of potential national and ethnic cultural differences can lead to serious cultural misunderstandings, which in turn can lead to ethnocentric value judgments and racism. Thus, the study of national cultural values and intercultural communication can be a powerful antidote to cultural misunderstandings and the racism that frequently follows from them.

Intercultural Communication

Some General Principles

The cultures surrounding the Mediterranean—Southern European, Near Eastern, and North African—as well as the cultures of Latin America, Africa, and the Middle East tend to be polychronic and high context (although the French have sometimes been labeled middle context). In contrast, the Germans, Scandinavians, British, and Dutch tend to be monochronic and low context. The European cultural roots of the original thirteen American colonies were largely British (or Dutch for the state of New York), which would tend to situate the dominant American communication pattern more toward the monochronic and low context end of the spectrum. However, the large number of Southern Europeans, Africans, Arabs, Latin Americans, Jews (depending on the origin point in Europe), and in some areas Native Americans would also tend to move Americans toward the polychronic and high context end of the spectrum, perhaps increasingly so in time. Certainly, there are huge regional differences within the United States, for example between Latin or African American communities in the Sunbelt states and Scandinavian or German ones in the Upper Midwest.

Using the findings of intercultural communication research requires thinking about culture in a flexible way that allows sweeping generalizations on the one hand and exceptions on the other hand. Sweeping terms such as "circum-Mediterranean cultures" and the ability to generalize from one culture to another are sometimes useful. At the same time generalizations about national cultures or culture areas have to be qualified to take into account regional, class, ethnic, and gender differences within a culture or region. Thus, the term *Latins* readily breaks down into any number of subdivisions such as lower-class men from rural Brazil or upper-middle-class women from San Juan.

Much of the research in intercultural communication is based on quantitative, social psychology studies. However, ethnographic studies that are rooted in long-term exposure to two or more cultures have the advantage of being able to get beyond the description of behavioral differences to the often unspoken and unconscious assumptions that lie behind those differences. An example of an ethnographic study of this type is *Cultural Misunderstandings: The French-American Experience* (in French, *Evidences invisibles*), by the cultural anthropologist Raymonde Carroll. A French woman who married an American man and has lived for many years in the United States, she experienced the everyday difficulties of life in a culture that—no matter how long one lives in it and no matter how well one speaks the language—constantly causes irritations because people do not behave in the way that she was brought up to expect. Rather than turn her frustrations into the usual stereotypes (Americans are materialistic; French are snobs), she developed a detailed set of case studies of "cultural misunderstandings" and a method patterned on psychoanalysis that she has labeled "cultural analysis."

Carroll's complex book provides a model of what sensitive cultural interpretation can bring to the problem of miscommunication across cultural boundaries. She studies a number of everyday situations, including the meanings of space in the home, relationships between parents and children, friendship, the use of the telephone and the etiquette of conversations, the sense of responsibility for minor accidents, and ways of obtaining information. In each of the cases she takes a simple misunderstanding and explains why the people reacted they way they did. She then uses the misunderstanding as the basis for an analysis that provides insight into each of the cultures.

Consider just one example to get a sense of Carroll's method. The example has some relevance for the topic of communication among scientists and intellectuals, for it involves a misunderstanding that occurred between an American and a French professor during a cocktail party. The French professor asked the American his opinion about the latest book of another American scholar. The American professor gave his answer, but during the course of the American's answer, the French professor stopped listening, started to look around the room, welcomed a newcomer who had just arrived, and interrupted the American's comments with a joke. The American felt that his French colleague had been insincere and really did not want to know his opinion. In turn, the Frenchman found the American to be a long-winded bore. Stereotypes were confirmed: the French are rude and Americans are bores.

Carroll investigates a number of similar cases by showing how cultural misunderstandings emerge from completely different assumptions about the way one is supposed to carry out the business of everyday life. In the case of conversational styles between two educated, white men, she likens the French conversation to a spider's web, with the exchange forming a tie or a bond between two people. In other words, the personal relationship supersedes the exchange of information. The conversation takes on a life of its own as a statement about the relationship. In contrast, she compares the American style of conversation to jazz or a jam session. The partners do not really have to know each other or have a relationship. The group makes room for the solo, and people take turns while they wait appreciatively through solo performances.

Although Carroll does not discuss the connection, the differing definitions of the conversation might be considered in terms of the individualism/holism distinction of Louis Dumont. The image of a jazz solo is highly individualistic, just as that of the spider's web is more holistic. In the individualistic American culture the social relationship—as symbolized by the conversational style—is considered secondary to the individual. For the French professor the conversation presupposes a social relationship; therefore, the content of what one says is, at least at times, secondary to keeping a lively, engaged conversation going, which is as much a commentary on the relationship as it is on the world.

Note also that Carroll's metaphor for the American conversation pattern is drawn from an African American music form. The jazz solo might be viewed in historical terms and compared with the African song form of solos and choruses. That musical form is rooted in traditional African society, which is polychronic and high context. Elsewhere Carroll argues that some of her findings about French-American cultural differences are similar to differences within the United States between Americans of Northern European and those of African descent. African American styles of communication are relatively polychronic and high context, at least relative to Americans of European descent, particularly Northern European Protestants.

Thus, there is an apparent contradiction: an art form of a relatively polychronic and high context ethnic culture is used as a symbol of the relatively monochronic and low context form of European-American conversation. One might question her choice of metaphors because it is not clear to what extent the jazz solo really symbolizes the dominant American conversational pattern. Jazz is, after all, an African American art form. However, one might also say that the jazz solo is a particularly appropriate metaphor because it shows how a musical form that may have one cultural meaning in one context may have an opposite one in another context. As an *African* American form, it is the legacy of the poly- chronic, African art form that also includes the chorus. As an African *American* art form, it dramatizes the value of the individual in American culture and conversations.

Yet another dimension that complicates the analysis is gender. In this case the American professor, when asked a question, responded by giv- ing a great deal of information. The French professor just wanted to chat; he did not want a dissertation. The difference between conversing to convey information and conversing to establish a relationship has some parallels with the differences in conversational styles between men and women. The conversation between the French and American professor was between two men. If the American professor had been a woman, would she have been better at reading the French male professor's cues and knowing when to stop taking up airspace?

The theories of the linguist Deborah Tannen (1990) suggest that the misunderstanding might not even have occurred had the American pro- fessor been a woman. Tannen argues that American men tend to engage in "report talk," whereas America women are more oriented toward "rap- port talk." In general, she finds that the male style of communication in the United States tends to be competitive and oriented toward locating people in a hierarchical relationship to each other. In contrast, she argues that women tend to be more oriented toward "connection" and less concerned with status hierarchies.

Note, however, that Tannen's use of the term *hierarchy* has a different meaning from the use for cross-cultural comparison. American men operate in a world that emphasizes achieved hierarchy, or social ordering, as an outcome of competition. In contrast, in Latin America, India, and other societies where traditional social structures are still strong, the concept of hierarchy is, at least in comparison, more ascribed or fixed. In other words, one is born into a certain position, and there is less room for social mobility than in a country such as the United States. Dumont argues that the one place where the United States is hierarchical in his sense of the term is in traditional race relations, which have sometimes been compared to the Indian caste system. Another area, which Dumont does not discuss, is arguably gender relations, at least in the traditional forms of patriarchal role assignments. Thus, in this sense I would argue that male report talk is part of a competitive form of hierarchy, whereas female rapport talk is related to the ascribed form of hierarchy associated with the subordinate position, which women often encounter in a relatively egalitarian society that remains, nonetheless, a patriarchy.

It is possible, then, that an American woman, with her emphasis on rapport and connection, would have been better at picking up the non-verbal cues of the French professor, and she probably would not have launched into a boring report, as the American male professor did. In this sense women's communication styles in the United States are not only rapport oriented, but relatively high context, and therefore more likely to be in synch with the communication style of the French and related cultures.

Body language involves yet a whole new set of possibilities for cultural misunderstandings. In Brazil, for instance, I learned that nearly everyone (regardless of age or sex) continually taps their conversation partner on the forearm to ensure their attention. The procedure seemed innocent enough between two men; it reminded me of a little child tapping a parent on the arm to maintain the parent's attention. It also seemed to be an expression of the general tendency that Brazilians have, like the French, to value personal relationships and one's position in a holistic social web. However, if a man from a Latin culture used this attention-getting device in a conversation with a woman from an Anglo culture (as I have seen happen), she might interpret his cue as an expression of unwanted sexual interest, and a misunderstanding might arise.

Differences in body language are perhaps most evident in the relations between Anglo Americans and Latin Americans. In the area of proxemics (spatial distance), for example, Anglos tend to keep a one-and-a-half to four-foot spatial bubble around them, whereas Latins keep an eight-inch to one-and-a-half-foot bubble. The differing bubbles result in the familiar situation of Latins backing Anglos up into walls during the course of a conversation. Likewise, in the area of haptics (touching), one study of a café in San Juan found that conversation partners touched at a rate of 180 times per hour, compared to 110 in Paris and 0 in London. The result is often a misunderstanding: Latins see Anglos as aloof, whereas Anglos may see Latins as aggressive or sexually interested. With handshakes, however, the relationship is sometimes the reverse: one study showed that Americans, who like strong handshakes, see Colombians as weak or untrustworthy, whereas Colombians see the Americans as excessively macho and superior-minded.

Sometimes the result of studies of this sort can be confusing. Occasionally, in Brazil I experienced vicelike handshakes by Brazilians who had probably read about American beliefs that a firm handshake is a sign of good character. However, in other cases studies of body language can be helpful. For example, Americans have a very casual attitude toward furniture. They tend to slouch in it, rearrange it, even put their feet up on it (usually only on their own furniture or that of close friends). To them, the behavior tells others not only that they are relaxed, friendly, and trustworthy, but also that they do not think they are any better than anyone else: they're just plain folks. It dramatizes the value of equality. Latin Americans find such behavior rude, and instead of dramatizing the value of equality, it may just show one's "lack of education" or impoliteness.

Intercultural Communication in the Workplace

Furniture is one area where Latin Americans converge with Germans, who are in many ways their cultural opposite. Consider a story narrated by Edward and Mildred Hall in *Understanding Cultural Differences* (83). An American working in the U.S. has an appointment to meet his new manager, who is from Germany. The American arrives on time (well, three minutes late) and knocks on his manager's door. The manager opens the door and, without smiling, motions the American to sit down. Without niceties, the German manager begins by giving a list of information he needs. The American, hoping to break the ice, moves his chair closer to the manager, who merely states that he needs the information by two o'clock that day and then rises from his chair to indicate that the meeting is over. Still hoping to break the ice, the American asks how the German's family is adjusting. The German says curtly that everything is fine and motions the American to leave.

Hall and Hall show that the American did just about everything wrong. He was late, which is intolerable for a subordinate in German culture. The American was trying to bring his spatial bubble closer to the German, who had a greater social distance, but by moving his furniture the American employee offended his manager. Likewise, the mixing of personal questions would have worked well for a French manager, but it is not considered appropriate for the Germans. To some extent, the American was more high context with respect to the German, but the misunderstanding also seems to be related to the American's attempt to flatten hierarchy and the German's desire to preserve it. In other words, both were operating out of their notions of proper social relations in the workplace; however, the two cultures emphasize the hierarchy/equality relationship differently.

A study of conversation patterns for German and American managers revealed even more profound differences (Friday 1989). Americans have a need to be liked. They are informal and group- or team-oriented, and they tend to use first names and "shoot the breeze" by talking about least-common denominator topics such as sports, the weather, current events, and families. Americans also tend to be assertive and direct when negotiating, but they have a sense of "British fair play." In other words, they tend not to "beat the opponent where they're down" and they tend to "put their cards on the table." In contrast, in business settings Germans have more of a need to be credible than to be liked. They are more formal and authority-oriented, and they tend to prefer to discuss the history of a topic and argue about its philosophical basis. Thus, Americans find Germans snobbish and authoritarian, whereas Germans find Americans shallow and uneducated. Like Americans, Germans are assertive and direct negotiators (in this way Americans and Germans are different from Latin Americans), but Germans tend to play "hard ball" or "go for blood," and they tend not to reveal their hand. The result is that possibilities for cultural misunderstandings are plentiful, even for the closely related cultures of Germans and Americans.

The case of German culture also shows that monochronic, low context cultures are not necessarily the most "modern" in terms of individualism and egalitarianism. In *Essays on Individualism* Dumont argues that historically German culture has been holistic and hierarchical relative to either France or Britain (and their descendent cultures). Dumont argues that historically in France a person was a human first and French as if by accident, whereas in Germany a person was human by virtue of being German first (130–31). In other words, in the German tradition one achieved a sense of being human through identification with the whole of German society or people (*Volk*). In contrast, in the French or Anglo-Saxon cultures one is first a human, a status that implies a host of associated values of natural rights and liberties. The state in France and Britain emerged as an institution that was to defend natural rights and liberties of individuals. In contrast, in Germany the state emerged as an expression of a social whole. The German concept of individualism has historically been linked less to individual liberties and rights than to the notion of *Bildung*, or self-education, as in the German literary tradition of the Bildungsroman, the novel of self-cultivation or personal development. Because for many centuries Germany was not unified and lacked a central state, the only way to achieve a sense of unity and wholeness was through personal cultivation and dedication to the arts and intellectual pursuits.

Dumont's analysis of German culture does not consider developments after World War II, and undoubtedly the long period of stable democracy in the former West Germany has led to a greater development of individualistic values that are closer to the historical French and Anglo-Saxon versions. (His analysis may now be more appropriate for the divide between the former East and West Germany.) Nevertheless, Dumont's analysis is interesting because it provides some insight into the intercultural communication studies that show why Germans tend to emphasize education and history in their discussions, or why they are, from an American's perspective, relatively authoritarian. The emphasis on history, as well as the tendency toward more authoritarian social forms, reflects the way in which the German concept of the self is historically more closely identified with the collectivity than in either the French and Anglo-Saxon cultural traditions. Likewise, the tendency to emphasize and value education, and to parade education in conversations in ways that Americans find merely snobbish, is related to the German definition of individualism as connected to the concept of "self-cultivation" (*Bildung*).

So far, the analysis has focused on distinctions within Western cultures. Much of what is said about Latin Americans and Southern Europeans will generalize, at least as a first approximation, to African and Moslem cultures. (There is, however, a strong streak of egalitarianism among devout Muslim men, one that even bridges the races, as Malcolm X learned in his travels to the Middle East.) The Asian cultures, particularly those of East Asia, present yet another set of possibilities. Consider, for example, the case of differences in communication styles between Americans and Japanese, which is now the topic of a huge literature and even a consulting industry. According to Hall and Hall (1987), Japanese tend to be very monochronic in their dealings with foreigners and technology; for example, they are very schedule-oriented when it comes to foreign business meetings or train schedules. In addition, they are very low context with foreigners; for example, they tend to ask lots of questions and require many details in business settings. However, in other circumstances, such as afterwork drinking sessions, the Japanese tend to be more polychronic. Furthermore, even within business settings the extreme importance of social position among the Japanese indicates that the high context nature of communication transverses situations.

As Hall and Hall point out and many business leaders have discovered, cultural misunderstandings between the Americans and the Japanese can represent substantial losses for businesses. For example, Hall and Hall discuss a negotiation in which an American company wanted to buy a Japanese company, and the Japanese company wanted to sell, but the deal went sour because the Americans wanted to get down to business immediately and the Japanese wanted to get to know their American buyers first (119–20). In another example, an American company sent an aggressive, young executive to Japan to manage its subsidiary. Within a few weeks he experienced management problems, and within six months he had lost his top Japanese manager (xvi–xvii). In a postmortem of the case Hall and Hall demonstrate the company's mistake by sending such a young manager, whose youth offended Japanese age hierarchies. The manager then compounded the mistake by adopting an aggressive style that undermined the organizational hierarchy of the company and affronted the more consensual style of Japanese decision making. Hall and Hall conclude their book with a list of advice to Americans collated from fifty American executives in Japan (145–51). The seasoned executives emphasize patience, "lying low" for about eighteen months to two years, taking time out for dinners, and above all learning the language. The last suggestion is, by the way, probably the number one rule of fieldwork for contemporary cultural anthropologists. If dreams are the royal road to the unconscious, so language is the royal road to communication through cultural difference.

Intercultural Communication Among Scientists

Because scientists (or engineers) are not just scientists but French, German, Brazilian, Japanese, or Canadian, the issues of intercultural communication discussed in the previous sections apply to scientists and other technical professionals. Unfortunately, there are few studies of intercultural communication differences or misunderstandings for the specific case of scientific and technical exchanges.

One starting point is the comparison of the style of scientific meetings, which the anthropologist Roberto Kant de Lima (1992) has pioneered, based on his observations of national anthropology meetings in Brazil and the United States. Kant notes that from the Latin perspective Americans appear rigid and inflexible (almost as the Japanese appear to Americans). For example, in the American anthropology meeting the entire schedule is made up months in advance, and people who present papers generally have to submit abstracts six months or more in advance. During the meeting there is a strict apportionment of times, so that everyone knows exactly when everyone else will be speaking. Usually the allotted time is the same for everyone except for a few selected plenary speakers. Kant reasons that this system reflects the Americans' preference for equality: the time limits for each talk ensure that everyone will get an equal voice, from the lowest graduate student to the most well-known chaired professor. Likewise, because every presentation has a unique time, it is possible to hop from one session to another, and thus everyone who is attending the conference is able to put together a unique itinerary for the conference and thus "individualize" the conference to suit his or her own tastes.

In contrast, Kant de Lima argues that in Brazil the program is much more flexible, so much so that "papers" may be accepted up until the last minute. Furthermore, the "papers" are often only oral presentations or talks; Brazilians tend to prefer oral communication because it is more flexible and personalized. (In my experience, Brazilians are wonderful conversationalists but terrible correspondents compared to Americans.) Furthermore, in Brazilian meetings the organization of time is much more flexible, and if a panel goes beyond its allotted time, it is a sign that the panel is more interesting to its organizers and participants. Thus, the length of time that speakers or panels take reflects their importance, and time is used to signal hierarchy among the speakers and panelists rather than equality.

Kant de Lima also discusses some interesting differences in conversational styles and the question-and-answer or discussion periods of conference sessions and other scientific lectures. In the United States, he notes, American academics and scientists tend to ask pointed questions. As long as the questions appear not to be motivated by personal animosities, then disputations and arguments—even what would appear to Brazilians as tactless or aggressive questioning—are accepted as part of the healthy dialogue of the scientific community. In contrast, in Brazil, to ask a polite question for more information may appear to be admitting that one does not understand the topic very well, and thus it would place the questioner in a relationship of relative inferiority with respect to the speaker. Likewise, to ask a more combative, challenging question might be interpreted as a break in friendship, an act of disloyalty, or an expression of a hidden agenda, such as allegiance to some other friend or network. Thus, in Brazil question-and-answer periods tend to degenerate into long declarations of principles, which can be interesting but highly veiled polemics that require a great deal of contextual knowledge to interpret. Brazilian question-and-answer periods may even shift into parallel conversations, a chaotic buzz of conversations that Americans would find to be a sign that order has given way to anarchy but that Brazilians might think of as a sign that the talk was so successful that it got people talking.

In short, when criticism occurs in Brazil, it tends to be more "between the lines," hidden in abstract and theoretical statements. In contrast, Americans tend to be more blunt and willing to disagree with their colleagues. It is difficult to explain why the cultural difference exists. There may be a historical aspect to it—Protestant sectarianism versus Catholic holism—but there are also more proximate factors, such as the legacy of the military dictatorship (1964–1985), which may still make Brazilians hesitant to criticize each other in public contexts.

One can see, then, the possibilities for cultural misunderstandings in the context of international scientific meetings. Brazilians who do not understand American communication patterns may find Americans blunt and tactless. In contrast, Americans might find Brazilians hard to pin down, elusive, overly theroretical, or too political. Considering the additional complexity of other international styles, international scientific meetings are a ripe territory for cultural (and other) misunderstandings.

In my experience the British are often extremely combative and polemical, a style that has much in common with their combative parliamentary sessions. Thus, even for cultural cousins such as the British and Anglophone North Americans there is a potential for grave cultural misunderstandings. For example, a British colleague once noted that in letters of recommendation it is standard practice to say something negative; otherwise, the letter may appear insincere. In American letters of recommendation, a negative comment may be interpreted as the tip of the iceberg, and one negative comment may be read as a crucial warning flag. Likewise, I have been told that junior professors in Britain see being criticized in print as a red badge of courage that shows their ideas are worth criticizing. In contrast, American junior faculty would fret over negative criticisms and book reviews as potentially dangerous for their careers. These differences of assumptions can lead to terrible misunderstandings when British scholars and scientists are asked to serve on review committees for junior Americans, often at the career expense of junior Americans. Furthermore, the cross-cultural differences can be magnified by gender differences, inasmuch as the polemical and openly critical style is sometimes associated with masculinist communication styles.

Differences among the various Western cultures pale in comparison to those between Western cultures and Asian cultures such as Japan. The anthropologist Sharon Traweek has spent more than fifteen years researching the cultural aspects of high-energy physicists. She has visited all the major research labs in Europe, the United States, and Japan, and she has completed extensive fieldwork in two labs: the Stanford Linear Accelerator (SLAC) and the Ko-Enerugie butsurigaku Kenkyusho (KEK), located in the new science city of Tsukuba, Japan. Much of her early research is summarized in the book *Beamtimes and Lifetimes* (1988), which includes a discussion of how physicists in both the United States and Japan have run up against cultural differences that belie their own ideology that the particle physics community is international and above cultural differences.

In one case the Japanese government made funds available that would help an American group purchase a detector in exchange for allowing a Japanese group access to their detector and laboratory (Traweek 1988:152–56). As the Japanese group visited the American group, its leaders came to think that it was necessary to state their plans and expectations in writing, a procedure that is not used in either American or Japanese labs. The Americans became extremely uncomfortable with the arrangement, and they asked the anthropologist for help in interpreting the Japanese request.

Apparently the Japanese group did not list the participating physicists in order of their reputations, and the Americans feared that the Japanese were planning to use the physicists in the order given on the list, which would suggest that the Japanese were not really serious about collaboration. However, Traweek showed the American physicists how the Japanese physicists were listed by the rank of their institutions, and within those categories by age. She noted that Americans list people by reputation and assume that people know the reputation of their institutions, whereas the Japanese list scientists by their age and institutional rank, and assume that readers know the reputation of the individual scientists. In other words, the American system highlights the individual position in an achieved hierarchy, whereas the Japanese system highlights the institutional and age hierarchy.

Traweek notes that as Americans and Japanese collaborate in the future, they will have to confront those and other potential cultural misunderstandings. In subsequent research (1992b) she has investigated how Japanese and American scientists have come to manipulate cultural stereotypes for their mutual benefit. For example, in a case where American scientists were working with Japanese scientists in Japan, the Japanese government made a decision that shut off resources to the laboratory. The Americans suggested to the Japanese scientists that they tell the government that the "American barbarians" were going to call a press conference and perhaps stage a demonstration. The Japanese passed on the word to the government, and funding was shortly forthcoming. By playing on stereotypes that the Japanese have of Americans as barbarians, both Japanese and American scientists were able to work together to achieve a goal to their mutual advantage.

The Comparative Social Organization of Scientific and Technical Communities

The potential for cultural misunderstandings is even greater when one considers how researchers bring with them different assumptions about the ways in which universities, laboratories, and related organizations should be and are structured. As I shall show in this section, there are tremendous cultural differences not only in the assumptions for every-day communication but also in those for the social structure of institutions. I will focus on university departments, although clearly there is room for much more research on the topic. University departments have been historically male-dominated, so for most of this section the discussion will focus on differences within these masculine social structures.

The United States

It may be helpful to begin with the United States, because the details of the social organization of academic research in this country may not be well-known. The dominant ideology of American culture emphasizes individualism and equality; hierarchy exists but in the ideal form it emerges as the result of a competition in which everyone is supposed to start from the same position and to enjoy the same rules. The pattern of the competition or race that encodes the values of individualism, self-reliance, equality, and achieved hierarchy can be seen in almost any domain of the culture. Metaphors of competition and self-reliance from baseball, football, and to some extent basketball percolate through the culture, and the country's economic ideology is based on principles of competition and free enterprise. Likewise, there is no royalty or aristocracy in the United States; instead, the European class hierarchy is replaced by the Hollywood stars, who are the successful ones in a hotly competitive world of entertainment. Americans also turn religious holidays such as Easter into a competition for the "golden egg." That race illuminates the early Protestant view that few would enter the gates of heaven; in a sense, the right to a place in heaven was earned through a competition that depended on one's ability to stay on the straight and narrow in this life. (The view extends to Catholicism, but the Calvinist form of Protestantism that shaped the dominant European culture in the United States did not allow for the middle ground of purgatory: you either won or you lost.)

I am describing the general "ideology" of the dominant European side of American culture. That ideology is obviously complicated by any number of countercultural currents, such as political and ethnic groups that emphasize noncompetitive relationships. Furthermore, in practice

American culture is considerably more hierarchical and less ideally competitive than it would portray itself to be ideally. For example, women have been historically excluded from many of the competitions and many domains of society, as have African Americans, Native Americans, and other underrepresented ethnic groups. The various forms of discrimination that are the topic of affirmative action lawsuits show how far the gap is between the reality and the ideology of competition where everyone begins the race in the same position. Nevertheless, those who have been excluded still use the ideology of universally equal rules and level playing fields to legitimate their struggles. Thus, ascribed hierarchy is seen as something "bad" that should be eliminated, and in this sense the United States is an egalitarian culture. Many of the debates, such as those over affirmative action, focus not on the end of reducing ascribed hierarchy but on the means for best reaching that end.

Science in the United States, especially academic science, participates in the general culture of, on the one hand, universalistic values that apply equally to individuals across contexts and, on the other hand, hierarchical ordering achieved through competitive individualism. For example, academic departments are usually based on a system known as collegial government (or government by colleagues). The department chair is generally democratically elected by the members of the department, and the chair achieves that position because he or she is seen as having special abilities as an adminstrator or leader. Although department chairs wield substantial power, they are usually voted into office and can be voted out by the members of the department. Frequently the position rotates among the different professors in the department, with the result that the position of department chair is sometimes reduced to an unwanted administrative post. The department members who are eligible to vote generally include all professors in the department who are tenured or who are untenured but in a tenure line. (Usually, assistant professors stay for six years before a tenure decision is taken. Then, the senior faculty either fire the assistant professors or vote to promote them to the rank of tenured associate professor, the intermediate step that precedes full professorship.) There are often other faculty members of the department—such as postdocs, lecturers, and adjunct professors—but they usually occupy only temporary positions and lack voting privileges.

In the American system young scientists who become assistant professors are immediately made colleagues with voting privileges, although some types of decisions are left to tenured faculty. Thus, the junior faculty are not placed under the wing of a senior faculty member or the department chair, at least in comparison with the relatively more hierarchical systems of other countries. The notion of tenure tracks or tenure lines is suggestive of an egalitarian system in which all faculty have a modicum of equality notwithstanding their rank as assistant, associate, or full professor. To use the metaphor of a competition, they have a lane (a track), and they are expected to achieve certain distances or cross hurdles within a given time period (see figure 5.1). Young scientists in the United States who are ambitious may jump from one institution to another one that offers them more prestige or a higher faculty rank. As in the corporate world, faculty are less locked into the departments in the United States than in other countries, and as individuals they are more able to move around from one institution to another.

American university departments and research laboratories function through a competitive system that is comparable to the star system in the worlds of sports, entertainment, or business management. When professors become famous because of their research or valuable because of their ability to win grants and awards (usually fame and value go together), they may sometimes receive offers from other universities. If the professors are fortunate, they can then play off the offers against one other and obtain privileges such as lower teaching loads, more funding and administrative assistance, higher discretionary budget, larger lab space and equipment, an increase in travel money, access to better graduate students, and the right to more leave time. At the top of the star system is the chaired professorship, that is, a professorship that usually comes with a title of a rich donor and perhaps some funding for conferences, research assistantships, and a secretary. However, chaired professors in the United States generally do not control their own independent institute; instead, they are usually in departments in which they will have the same one-faculty-one-vote privileges as other faculty.

The star system of many American organizations has some similarities to the structure of the traditional American family in European-descent populations. That family is nuclear rather than extended, although of course in recent years a number of variants of the nuclear family have emerged (such as single-parent families, unmarried couples of various sexual preferences, and nuclear families with children from previous marriages). The traditional American nuclear family is also generally neolocal in the sense that children usually leave the home ("cut the umbilical cord") when they grow up. In a sense children exist

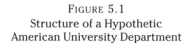

FIGURE 5.1
Structure of a Hypothetic
American University Department

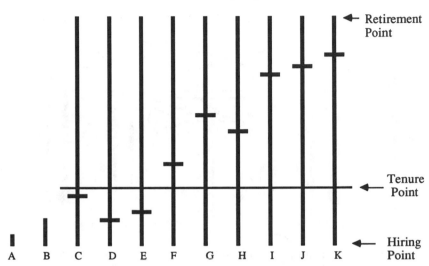

Key: Professor A is an adjunct professor, with a one-year appointment, whereas B is a three-year visiting assistant professor. C through D are all hired on lifetime tenure-stream appointments. C through E are not yet tenured; they may be fired when they come up for tenure in their sixth year. F through H are at the associate level, and I through K are at the full professor level. Generally, one of the full professors also serves as the department chair. The short horizontal bar indicates the number of years served since the hiring point. (However, it is possible to be hired at a higher rank.)

in "lines" as individuals who are encouraged early to develop their own talents and their own special path through life. When they reach adulthood, they leave the nest. As they grow up, children learn that their achievements make their parents "proud," and in a competitive culture comparisons among siblings, schoolmates, and cousins—no matter how much they are muted by a discourse of egalitarian parental love—are inevitable. Although comparisons among children and sibling rivalry are by no means unique to the American family, they have a special meaning in a culture based on individual competition and achievement. American families therefore produce their own sort of star system, in which the successful child is compared explicitly or implicitly with the less successful one. In family reunions the size of one's car and house becomes a readily visible and easily quantifiable measure of success for

FIGURE 5.2

Hiring Patterns Among American Anthropology Departments

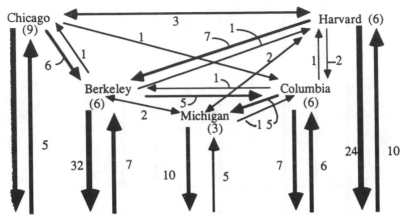

Group of next largest Ph.D producers: Arizona, Cornell, Illinois, Indiana, Pennsylvania, Stanford, UCLA, Wisconsin, Yale

Source: List of departments by member and by number of Ph.D.'s granted to people in academic positions, from the American Anthropological Association Guide to Departments, 1992-93. The top five departments have all produced more than 200 Ph.D.'s who hold academic positions; the next group has produced 100–150 Ph.D.'s, with no group in between. The number in parentheses are numbers of "home-grown" Ph.D.'s.

everyone to see. Stars may be born, but along with them come the family failures.

The competitive relations among colleagues of the same generation therefore can be compared with sibling rivalry and other forms of peer competition that American children learn as part of their childhood socialization. A second level of competition emerges among the scientific and academic departments, which in the United States are ranked based on surveys of leaders in the field. Unlike the Japanese system, the ranked hierarchy of departments is much more achieved rather than ascribed; that is, the departmental hierarchy in the United States is capable of relatively rapid changes. Thus, the order is more like the ranking of sports teams at the end of each season, and as players move from one team/department to another, its reputation may go up or down. The flexibility of the hierarchy is important, because the ranking-type hierarchy emerges from principles of equality and competition.

A few departments are major Ph.D. producers, and they tend to be the departments with the highest reputation. For example, in the field of anthropology, five university departments have each produced more than 200 Ph.D.'s who hold full-time academic faculty positions (see figure 5.2). Another nine departments have produced between 100 and 150 academic Ph.D.'s each, and there is no middle group of departments between 150 and 200. The fourteen departments that are responsible for over 100 Ph.D.'s each are also responsible for a very large percentage of the total number of Ph.D.'s in academic positions in the United States, and the thirteen departments also hire a majority of their own faculty from each other. Thus, a fairly close-knit exchange system defines the elite of the academic departments.

In some traditional societies clans that are "spouse givers" (often lineages governed by men whose sisters and daughters marry into other lineages) see themselves as having higher status than the "spouse takers." In a similar way, the "Ph.D. givers" (especially Chicago, Harvard, and Berkeley in the case of anthropologists) are more highly ranked than the "Ph.D. receivers." Furthermore, in some ranked societies the top clan tends to seek wives from outside the village or the tribe. In a similar way, the top three schools also have a high percentage of faculty who earned their Ph.D.'s from foreign universities, especially from Britain. It is as if the Ph.D.'s from schools below them in the ranking are not good enough, and perhaps there is a residual sense of colonialist thinking involved in the idea that anthropologists trained in Britain or France may be more sophisticated than those trained in the United States. (The opposite may be the case in the natural sciences.) Finally, in some societies, royal lineages have been known to marry within the family and therefore violate the nearly universal incest taboo. In a similar way, the highest ranked departments tend to hire more of their own Ph.D.'s, probably because they consider their own Ph.D.'s to be the best. (The number of "homegrown" Ph.D.'s is listed in parentheses in figure 5.2.)

The system of ranked exchanges through which the academic discipline reproduces itself is sometimes referred to as the *old boys' network*. As more women achieve positions in the academy, a more appropriate term might be the *old school network*. However, notwithstanding increased workforce diversity, such networks are likely to survive in a modified form. For example, in anthropology departments of historically women's schools where there are a large number of women professors, such as Vassar and Wellesley, the sources of the Ph.D.'s in my sample are from elite institutions. The old school network is an instance in which the relative equality and individualism of American culture is mitigated by notions of departmental hierarchy and personal ties. However, even in this context notions of hierarchy and personalism are rooted in cultural values of equality and individualism. For example, old school networks are legitimated by the assumption that the Ph.D.'s from top-ranked departments have earned a position in those departments due to prior competitive ordering. Furthermore, operating against this system of departmental ranking is the star system, which provides for social mobility and injects individualist competition into the system. Thus, very productive professors who have top-notch reputations, even if their Ph.D.'s are from lower status schools, may eventually get offers from top-ranked departments, especially if they have built solid professional networks over the course of their careers.

One aspect of reputation is the number of students that an academic scientist has produced. To produce the best students it helps to be in a top-ranked department, which has the funding and the reputation to attract them. When professors move from one institution to another, they sometimes take their graduate students with them. However, when the student earns the Ph.D., it is generally considered a sign of healthy independence (and the strength of one's mentors' old school ties) for the student to move on to a position in a different institution. As noted, the exception is at the top of the hierarchy of ranked institutions, where departments tend to hire more from their own graduate students. There is also a tendency for low-prestige departments to hire their own graduates. Presumably that is because high-prestige departments think their own students are the best, and low-prestige departments have trouble placing their students or attracting suitable faculty.

In anthropology and other disciplines students become identified closely with their Ph.D. advisor. In turn, the advisor is often identified as the student of someone else in an earlier generation. Thus, people tend to think in terms of scientific lineages, which are based on a notion of descent through a chain of advisors. These "virtual lineages" are another, more concrete form in which old school networks operate. In anthropology, for example, the students of the Columbia anthropologist Franz Boas founded the anthropology department at Berkeley, and a large number of anthropologists in other universities trace their lineage to Boas. Likewise, another group of anthropologists and sociologists trace their lineage to the Harvard sociologist Talcott Parsons, whose students included Robert Bellah, Clifford Geertz, and David Schneider. The historian Donna Haraway has also mapped out an important lineage in physical anthropology and primatology around the students of Sherwood Washburn (1989).

In general, the idea of virtual or scientific lineages seems to be a very archaic or hierarchical idea at the heart of a modern, individualistic, and egalitarian scientific culture. However, clanlike membership in a viritual lineage is not enough to guarantee a student's future. A WASP, male, upper-class student from a top-ranked school who studied under a highly respected advisor will have doors opened for him. Still, advisors will not spend their symbolic capital or risk their reputation on a poor student, and even students from top-ranked programs are usually "downwardly mobile" (that is, they start off teaching in lower status schools). Although students with privileged starting points usually have a head start, they still have to earn their way back into the top-ranked schools based on their reputation for research, collegiality, and teaching. Likewise, if they are lucky enough and well-connected enough to be hired directly into a top-ranked school, they still have to earn tenure there. As Traweek notes, in the United States members of each new generation of scientists must construct their networks and resources all over again (1988:151).

In general, the advisor/student relationship in the United States is a relatively weak bond. Once students are established in a new department, and especially after they have tenure, they often take off on a new path of research that is quite independent from that of their advisor. For example, as discussed in chapter 2, Haraway (1989) shows how women who belonged to the Washburn lineage eventually were able to call into question the sexist bias of some of their advisor's theories of primate behavior. Furthermore, in order for students to rise and establish reputations, they need to exercise a certain amount of independence from their advisors. To work too much in their advisor's shadow cuts against the grain of individualism, which assumes that students who work too closely to their advisor's line of research may not have the suitable brilliance and originality to become stars. Thus, students face a delicate balancing act between loyalty to their advisor and individualistic independence. As a result, they build networks that include people and resources from their advisor's networks, which are grafted onto their own networks, which in turn have a strong component of same-generational ties. From my own observations of colleagues I suspect that the growth of E-mail and virtual communities will only further weaken the importance of the advisor/student bond and old school ties, and in turn it will strengthen the horizontal ties among same-generation networks that young colleagues must build when they leave the nest of graduate school.

Europe

In contrast to the American university system, many of the other university systems in the world have a series of features that can be summarized as relatively hierarchical. Americans with little foreign experience often fail to realize that theirs is the odd culture. In Europe, and in other continents where universities are based on the traditional European model, collegial governance is sometimes replaced by a system of powerful full professors who control a department or institute that has a number of junior positions. Although there is a great deal of variety among the European forms of university departmental structure, this section will focus on the traditional departmental structure in which power is centered on a professor who in turn controls an institute or some unit that includes subordinate positions. That structure is still in effect in many German universities today, although the system is also modified in significant ways. For example, reforms during the last decades of the twentieth century have allowed for multiple professors in some institutes.

From an American's perspective the power of full professors may appear to be a twentieth-century survival of European aristocratic privilege in which professors control minifiefdoms. Likewise, Americans might see the European structure as a translation of more authoritarian and patriarchal family structures. Although the classical European form may appear premodern or undemocratic from an American's perspective, the communication style of many Europeans is paradoxically more democratic, at least to the extent that open disagreement and debate is considered more democratic. In my experience, Europeans, at least Northern Europeans, tend to be more openly critical of their colleagues and more prone to statements that to North Americans seem aggressively critical. (In turn, as Roberto Kant de Lima noted, North Americans are more openly critical and seemingly polemical than Latin Americans.)

Although different communication patterns are part of a much larger system of macrosociological differences, they are also products of, just as they reproduce, differences in local academic social structures. For example, Americans are only granted tenure after six years of serving as assistant professors. Consequently, they are much more vulnerable to criticism than junior researchers in other countries, who earn the equivalent of tenure earlier in their careers. (In Britain the Thatcher government abolished tenure, and it remains to be seen if this change will lead to greater mobility and, with it, a change of communication styles.) In the American system and other places where vertical and horizontal mobility are closely connected, severe criticism in print may affect the reputation of junior and mid-career scientists in ways that affect their ability to hop across institutions and therefore to move up the academic hierarchy. Thus, in the American system one tends to find polemics directed upward toward the senior members of the academic/scientific hierarchy or across among senior members. However, in my experience junior-to-junior or senior-to-junior polemics are more likely to be seen as motivated by personal reasons. Europeans who do not understand the dynamics and variable vulnerability of critique in the American system may therefore engender very hard feelings from their American colleagues without even understanding exactly why those feelings exist in the first place.

In addition to differences in communication styles, researchers have also begun to unravel how university social structure is related to national styles in the content of scientific research: methods, theories, and a sense of what research is important. Some historians and sociologists of science have assumed that national styles in science, at least among First World countries, largely disappeared by the twentieth century. However, as recent historical research and some of the cases discussed in chapter 2 have shown, national styles are more resilient than may have first been thought, and those national styles often coincide with differences in the social structure of research organizations.

In *Styles of Scientific Thought* the historian Jonathan Harwood examines comparatively genetics research in Germany and the United States during the modernist period (specifically, 1900 to 1933). American geneticists during that period tended to emphasize specialized and empirical research with a pragmatic orientation, whereas Germans of that period were more theoretical and concerned with broader questions of evolutionary theory and development. Certainly the broad cultural legacy of empiricism, as I have already discussed for early modern English culture, was a contributing factor to the American style, but Harwood's analysis shows how the pragmatic and empirical style of the United States is reproduced through a specific social organization of university research. For example, by the early twentieth century American universities were already under a collegial form of departmental governance. In other words, junior faculty were hired into tenurable lines and were therefore relatively independent of the senior faculty. Although American universities had stronger presidents than their German counterparts, American department chairs were often elected by faculty members. Another distinguishing feature of American departments is that faculty pay was not based on the number of students taught. The result was that in the United States senior professors tended to leave large introductory courses to junior faculty, and the senior faculty opted for small specialized courses. Finally, in the United States genetics departments were often housed in agriculture schools or funded through state legislatures and foundations interested in practical applications. Those factors contributed to a tendency for American genetics researchers to specialize and to develop a narrowly focused, empirical style of research, in contrast to the Germans' preference for a genetic science that could explain the role of genes in a broader range of phenomena.

Harwood shows how the German style was, like the American style, shaped at least in part by the institutional context of early twentieth-century German universities. Unlike the American system of tenurable lines, in Germany junior faculty worked either as a *Privatdozente*, which could be compared to an adjunct position in the United States, or as an *Extraordinarius*, which was usually a nontenured term appointment equivalent to, say, a five-year contract as an assistant professor in the United States. Thus, no system of parallel tenure tracks existed in which faculty moved up from assistant, to associate, to full professor. In Germany hiring and funding decisions were also relatively more centralized in the hands of the full professor (*Ordinarius*), who was often the chair of an institute. Although comparable to American department chairs or chaired full professors, the institute chairs were more powerful because they were appointed by the state-level ministry of culture rather than elected by the faculty in the department. Thus, the institute chairs were able to make important decisions without the accountability that was and is typical in American universities. Institute chairs were able—and in many cases today still are able—to negotiate directly with the ministry all salaries, pay for staff, office supplies, and books. Furthermore, because the chairs were appointed directly by the government, the institute chairs also tended to have more power than American department chairs with respect to the university president or rector.

A related structural difference involved the system of salaries and fees in Germany. Wages were to some extent based on the number of students taught, and consequently senior faculty tended to teach large introductory courses, which required a generalist's knowledge. Because there were fewer chairs and there was less departmental specialization, junior faculty had a career advantage if they maintained a breadth of research interests. As a result, when they reached a rank of *Extraordinarius,* they were in a better position to compete for the few professorships as they opened up.

The theoretical and general orientation of the German geneticists was therefore connected to the relatively hierarchical structure of the academy. However, Harwood also notes that there were many cross-currents that complicated the overall difference in styles and social structures between American and German geneticists of the time. For example, within Germany there was a division between two groups: the "mandarin" type of professors with educated, middle-class backgrounds who had a more comprehensive style, and the "outsider" group with less elite backgrounds and a more pragmatic, "American" style. There is also some evidence that a similar difference operated within American genetics at the time, although between the elite, private universities and the agricultural, state universities. Furthermore, after Hitler came to power and Jews were fired from the universities, the German situation became complicated by more explicitly ideological divisions. German geneticists' responses to the Nazis were very diverse, and ultimately many Germans emigrated to the United States, thus, further complicating the question of national styles.

Harwood's analysis is instructive for reasons that go beyond a demonstration of the existence of national and local style variations within a scientific community. By focusing on comparative social organization, he shows how it is possible to avoid simplistic formulas that explain differences in national styles of science by reference to vague concepts such as national cultural traditions. Rather, those concepts should be seen as rubrics that point to complex variations in ongoing institutional structures. Although it might be tempting to see the differences in institutional structures as the cause and the differences in research or communication styles as the effect, it would be more accurate to say that the patterns are coproduced. The social structure of science in any given country may strengthen a particular scientific research style, but the scientists in that country are likely to value that style and to see their style as a point of reference in struggles over attempts to reform or change the social structure. In other words, the two go together in a process of ongoing mutual reproduction and transformation that shapes and is shaped by the people involved.

Japan

On the surface the social structure of Japanese university science is similar to that of traditional European universities. To some extent, the similarities are due to explicit borrowing. For example, according to the historian James Bartholomew large numbers of Japanese scientists during the first decades of the twentieth century studied in Germany (1989). Notwithstanding the many influences, the Japanese did not completely copy the German departmental and institute structure. Rather, the particular form of hierarchy that was instituted in Japanese university structures was rooted in a very different cultural tradition and society. My discussion of Japanese science will therefore first consider some general background studies on Japanese culture.

In *Japanese Society* the anthropologist Chie Nakane develops a detailed analysis of the Japanese form of hierarchy in social structure. She shows how traditional villages, modern corporations, political parties, families, and other social organizations in Japan all tend to be organized along the principle of an inverted *V*. In other words, groups attach themselves not through horizontal bonds but through vertical allegiances to a higher status group or leader. The structure is replicated in a long chain of bifurcating organizational connections, with bonds running vertically rather than horizontally (see figure 5.3). Nakane compares this structure to what anthropologists call "segmentary lineages," that is, a vertical system of social organization in which loyalties run along figurative "parent-child" bonds rather than across in "aunt-niece," "sibling," or "cousin" bonds. In the Japanese system it is relatively difficult to establish horizontal bonds such as a national trade union movement that would cut across company unions.

The basic vertical structure of Japanese organizations is found in most social organizations in Japan, including the Japanese family or household unit, the *ie*. In American and other Western families, kinship is "bilateral"; that is, relatives are counted on both their mother's and father's sides.

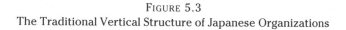

FIGURE 5.3
The Traditional Vertical Structure of Japanese Organizations

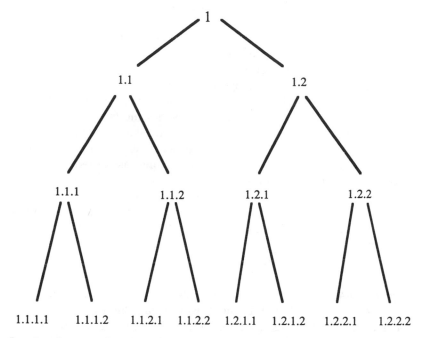

Based on discussions in Nakane (1970)

Kinship bonds extend outward from the ego in all directions: vertically down to children and grandchildren and up to parents and grandparents, but also horizontally to aunts, uncles, nephews, nieces, cousins, and second cousins. Furthermore, when Americans marry or reach adulthood, they usually set up a new household and achieve independence from their household of origin. As a result, family ties—both marital bonds and the continued ties with one's family of origin—are based to a large degree on choice. Americans will sometimes even cut off relations with parents, siblings, or other relatives whom they do not like, and certainly Americans have a great deal of control over the quantity and quality of the ties they have with their relatives. In other words, the American family is an extremely fluid institution, and the strength and intensity of the familial bond depends to a large extent on choice. A notion of social contract among individuals therefore permeates the American family.

In Japan the family is patrilineal; that is, the child belongs to the father's clan. It is also patrilocal; that is, after marriage the wife frequently goes to live with her husband's family. Usually when the oldest male relative in a family dies, the oldest son takes over as the head of the family. The rule is known as primogeniture, and one can find evidence of it in Western cultures dating back at least to the book of Genesis. If the oldest son is not competent, the job sometimes passes to another son, and if there is no son available, the daughter usually remains in the family and a son-in-law is recruited from the outside. Thus, the positions in the family structure always remain, but the persons may move within them.

Nakane notes that the vertical structure of the Japanese family is similar to that found in business organizations, and the anthropologist Matthews Masayuki Hamabata found that in some cases there is even an overlap between the *ie* and the vertical structure of the business enterprise. For example, if the head of the family owns a business, the various sons might become heads of subsidiaries. In these families the hierarchical structure of the Japanese patrilineal family becomes the backbone of and model for the Japanese business empire. He sums up the *ie* as follows:

> The *ie*, therefore, is a perfect example of sociocentric forms of organizational life [what Dumont would call "holism"]. Simply because it is not made up of individuals but of positions, the *ie*'s existence, unlike that of the nuclear family's or the extended family's, is highly resistant to the unpredictability of individual characteristics and behavior. (1990:40)

The vertical structural principle that Nakane, Hamabata, and others elucidate can be seen in other cultural domains, such as Japanese preferences for organizing the religious world. For example, Buddhism and Shintoism support beliefs and practices oriented toward veneration of the dead, Confucianist doctrines support filial piety, and households frequently have altars dedicated to the dead. The Shinto rituals associated with the ancestors and other spirits dramatize a spirit world that, like the social world, is imbued with a sense of hierarchy, respect, and duty. As the sociologist Robert Bellah describes it, the category of supernatural beings "shades off imperceptibly into political superiors and parents, both of whom are treated as in part, at least, sacred" (1957:61).

Bringing up the example of religion can be useful to dispel associations that Westerners may have with the word *hierarchy,* given the Western predilection for equality. Westerners tend to gloss vertical hierarchy too easily as authoritarian, and as I shall show in the next chapter hierarchical institutions may indeed become authoritarian in a Western context. However, Japanese hierarchy involves more than a sense of debt and obligation from juniors; it involves all sorts of responsibilities from the seniors: to maintain a close-knit organization, to move over when their time comes to retire, to create a nurturing environment for the juniors, and to consult with the juniors and attempt to reach a consensus when making decisions. In other words, Japanese hierarchy is negotiated and complex. As Nakane describes it, the ideal sense of hierarchy in Japan is like that of a well-functioning parent-child relationship, which is the metaphor that Japanese often use for their organizational bonds. For Westerners the metaphor of a parent-child relationship again might have overtones of unbearable paternalism. It might be better, then, to think of Japanese hierarchy in terms of the vertical relationship between the believer and a benevolent god or a guardian spirit, in which obligations flow both ways.

Given this background, the vertical organization of scientific research groups in Japan becomes more comprehensible. In *Beamtimes and Lifetimes* Traweek discusses the organizational structure of Japanese university research groups, which provide the academic home or the organizational model for most high-energy physicists and their laboratories. The Japanese system is considerably different from its American counterpart. As in the German system described by Harwood, departments are built around a few full professors who function as heads of mini-empires. As Traweek (97) explains, in Japan government funding goes not to the university or laboratory but to the *koza*, or group led by a full professor (*kyoju*). The group generally includes one associate professor (*jokyoju*), one research associate (*koshi*), and two assistants (*joshu*).

In Japan, while there is equality among the full professors in the broader context of university governance, the junior people are much more under the thumb of the senior faculty than in the American system. Unlike junior faculty in American universities—that is, the nontenured assistant professors—the Japanese associate professor and research associate do have tenure. As a result, when the full professor retires the associate professor generally moves up, and the research associate often becomes an associate professor. This transition of ranks is comparable to the transition of ranks in the Japanese family when the patriarch dies: a younger male becomes the head of the household, and the other males also move one step closer to the status of head of household or they move out and start their own household.

In some cases, particularly during the earlier parts of the twentieth century, the relationship between the Japanese department and *ie* was even closer. Bartholomew shows that there was a great deal of intermarriage between and within university departments. Some professors actually chose their successor by marrying their daughters to their favored student. That practice is somewhat different from the *ie*, in which the eldest son moves into the patriarch's position and marries a woman from outside the family. However, the practice is similar to that of families without sons, so there is a precedent for it in the family structure. Furthermore, the practice of choosing a successor as a son-in-law provides the system with the flexibility that is needed for a profession in which competence is important.

In both the *koza* and the *ie* there is room for negotiation, especially if the person next in line is not particularly competent for the position. However, the basic model underlying the system remains a structure of hierarchy that contrasts with the more competitive system of the United States. To review, in the American system each professor, regardless of rank, occupies a line. When the line is vacated, the department forms a search committee to try to find someone from the outside to replace the professor who has left. Sometimes a junior person is hired in the line, and when the department is lucky enough to convince the dean or president for more funding, the department may hire someone more senior. However, when a senior person vacates a position in the United States, there is no overall shift in positions within the structure.

The result is that in the Japanese system the junior faculty have considerably more job security than in the American system. However, at the same time the Japanese researchers are more locked into a *koza*, or research group, and there is less opportunity to move from one *koza* to another. The Japanese research system, then, is very similar to the traditional employment system of Japanese corporations, where the worker exchanges lifetime job security for lifetime loyalty. The *koza* system is also quite different from the star system of American corporations and universities, where executives/professors hop from company to company or university to university on their way up the ladder. Because of the bifurcating vertical structure in Japan, it is difficult to move across organizations.

The overall implications of the system for the Japanese is more security but less mobility. Whereas the United States is a culture of rising stars, Japan is a culture of rising suns (or sons). The Japanese system is relatively more hierarchical and holistic, because scientists are immediately assigned a position in a *koza*, and they are locked into the position. The *koza* leader also passes on to the junior members a legacy of funding, instruments, and ties to his successor. What survives is the *koza*, which as a corporate group like a patrilineal family takes on a reality of its own and in some ways is the more fundamental social unit than the individual researchers who occupy positions in it (Traweek 1988:151).

Although the Japanese social organization of academic science operates according to hierarchical and holistic principles, at least in comparison with the United States, there is a paradox: Traweek notes that the Japanese scientists see themselves as more democratic than their American counterparts (147). Within a research group in Japan there is a tradition of interdependence (*amae*), and the group makes decisions according to consensus. Thus, Japanese group leaders must fully consult with members of the group before making a decision. In contrast, in American physics laboratories the group leader makes major decisions, and the role of group members, such as graduate students and technicians, is more to implement rather than to generate new ideas.

However, the paradox is resolved when one considers that the Japanese research group includes what would be the equivalent of assistant and associate professors in the United States. Thus, it would not be exactly parallel to compare the collegial relations among the Japanese full professor, associate professor, and research associates with the more hierarchical relationship between the American group leader and laboratory employees. Instead, in the United States people of the rank equivalent to the Japanese associate professor or research associate (i.e., assistant professor) are more likely to be doing their own research or to have formed partnership relations with their senior colleagues. From this perspective, the American system is more egalitarian and democratic. As in the case of the European/American university comparisons, issues such as who is more democratic or egalitarian depend on what units one selects for comparison and how one defines such fuzzy terms. The moral here is a cautionary tale against rejecting other social forms as inherently authoritarian or undemocratic. A better approach is to locate where authoritarianism and democracy operate at different points within each system's structure.

Traweek argues that both Japanese and Americans borrow on social organizational models from their surrounding culture. (This could be seen as a type of social-to-social totemism.) In the case of the Japanese, the model is the household (*ie*), which is hierarchically organized according to age and gender (148). Decisions are made after the head of the household consults everyone, but ultimately the decision-making power rests in the head of household, who is male (because women have much less power relative to American households). In contrast, in the U.S. the cultural model for physics research groups is more likely to be the sports team, in which the coach has authority but ultimately the very existence of the team is contingent on the players' consent to be coached, because players can leave the team if they are not happy. As Traweek notes, the team model "generates very strong professional loyalties and very weak institutional ties"—the opposite of the Japanese model (149). In one way, then, the Japanese system is more democratic—because there is more consultation—but in another way the American system is more democratic—because ultimately the players can take their balls, go home, and find another team. The American model, then, rests on the principle of consent by the governed, a principle that stretches back not only to the American Constitution and Mayflower Contract but also to the English tradition of parliamentary democracy and the Magna Carta. In a similar way, Nakane notes that the vertical structure of Japanese organizations is rooted in the feudal village structure but reproduced in modern corporations. In other words, these organizational principles are at once very contemporary and very archaic— they are cultural structures.

Comparisons at such a broad level are, of course, subject to many exceptions, and the hazard of doing cultural comparison across national cultures is to assume that there are no differences within the cultures. Harwood underscores the point in his careful study of internal differences within the early twentieth-century German genetics community. As anyone who has traveled through the various regions of the United States or Japan knows, there are also impressive variations across local settings. On this point, Traweek notes that "*both* models exists in *both* countries," and thus generalizations must always be interpreted cautiously for what they are: generalizations. Nevertheless, the generalizations provide insight into significant differences, and they serve as useful guideposts for social navigation in a transnational world rife with potential for cultural misunderstandings.

Brazil and Mexico

An American worker in a Japanese-owned automobile factory once complained about the Japanese: "The biggest difference between us and the Japanese is that we work to live and [they] live to work" (Fucini and Fucini 106). The quote is interesting, because sometimes Latin Americans describe the difference between themselves and North Americans in identical terms. This section of the chapter moves from an East/West dimension to a North/South dimension, using Brazil and Mexico as the case studies.

Latin America may be considered "Western" because its official organizational structures are inherited from Europe, the dominant religion is generally Christianity, and in most places the dominant language is either Spanish and Portuguese. (In fact, use of those languages is sometimes a distinguishing feature for the definition of "Latin America.") However, Latin America is also the product of one aspect of Western culture: early modern Spain and Portugal. The settlers from the Iberian peninsula generally came to establish large plantations or mining operations rather than to set up small family farms as in the northern United States and in parts of Canada. There is, then, a sharp difference in classes in Latin America, which corresponds to racial and ethnic differences. The privileged members of the middle and upper classes are predominantly of European descent, and the descendents of the Native Americans and former African slaves continue to occupy the more menial jobs and positions in the society. The idea that Latins "work to live" reflects, in part, the aristocratic legacy of the plantation and the colonial lifestyle. "Work" belonged to the lower classes, and it was generally not very meaningful, whereas leisure activities ("mental work") was the province of the upper classes.

Latin American societies therefore are—in comparison with those of Western Europe and North America (that is, the United States and Canada)—relatively hierarchical. However, the principle of hierarchy is different from the replicating inverted *V* structure that Nakane described for Japan. Not only are horizontal bonds more important, but power tends to centralize at the top in the figures that are so famous in Latin American culture: the strong man (*caudillo*), the pope, the bishop, the dictator, the *patrón*, or the large landowner (*latifundiário*). The centralization of power leads to resistance via horizontal bonds that appear in the form of democratizing or modernizing movements, and as a result Latin American history tends to oscillate between centralizing and decentralizing movements of authoritarian repression and revolutionary anarchy.

One way into the structure of Latin American cultures is through the Latin American family, which is distinct from either the Japanese *ie* or the American nuclear family (and its variants). Unlike the Japanese *ie*, the Latin American family is Western in that it is bilateral and neolocal. In other words, children count as relatives both their father's and mother's families, and when the children grow up and get married, they eventually form their own families (although adult children tend to remain at home longer and live closer to home than in the United States). Unlike the nuclear American family, the Latin American family is a huge, extended network—people can sometimes name more than two hundred relatives—in which the father often rules as a patriarch and for which friends are turned into family members rather than vice versa, as tends to be the case in the United States. Family ties readily incorporate friendships, so that some people are able to build huge personal networks of horizontal and vertical bonds. Through combinations of horizontal linkages to one's relatives, friends, and their vertical linkages to the older and more powerful, the less privileged can sometimes garner jobs and favors in a system that is both authoritarian and flexible.

Similar relationships appear in Latin American Catholicism. In Calvinistic Protestantism the relationship between the self and God was dyadic and unmediated, whereas in Roman Catholicism there was always a third party: the priest, the church hierarchy, or one's patron saint. The Catholic church was (and still is) a hierarchical organization; it rests on a great hierarchical chain that runs from God, through Jesus, to Peter (the "rock" upon which the Church was built), through the pope, to the archbishops and bishops, down to the priests, and, finally, to the believer. In contrast, the Protestant church was defined as a community of true believers who came together as equals and formed a congregation. In that structure the minister or pastor was chosen by the congregation to lead them but not to mediate between them and God. In this sense, Latin American Catholicism appears to be structured according to something like the Japanese principle of vertical hierarchy. However, in Latin American Catholicism there are crosscutting structures that mitigate the hierarchy: religious orders and fraternities, an ideology of the equality of souls before God, and popular religious traditions that allow intervention in the religious hierarchy with the help of protective spirits, saints, or Virgin Mary.

The traditional Latin American state and agricultural organizations operated according to similar hierarchical principles. Unlike the constitutional monarchies that emerged in early modern Britain and Holland, in Spain and Portugal absolute monarchies solidified, in no small measure due to the gold that flowed into their coffers from the New World. As a result, authority centralized on the monarch as the head of the state bureaucracy. The monarchs granted large concessions that led to the *latifundia* organization of agriculture, that is, organization into large plantations or ranches headed by a patriarch who held authority over the people in his dominion. The result was a series of quarreling fiefdoms in which rural bosses formed factional alliances against each other. Vertical bonds took the form of patron-client ties, in which the subservient exchanged loyalty for protection. Just as the Catholic believer could achieve some flexibility in the religious hierarchy through the intervention of patron saints or spirit guides, so the adept poor person could manipulate the social hierarchy through patron-client ties and extended kinship networks.

To summarize, unlike the replicating vertical hierarchy of Japan or the competitive achieved hierarchy of the United States, in Latin America there is what one might call a mediated hierarchy, that is, a very flexible system that permits so many exceptions that the system tends to oscillate between extremes of authoritarianism and anarchy. The mediated hierarchy also allows intermediaries (such as saints or rich patrons) to intervene and bend the rules in favor of someone who is lower in the hierarchy. The importance of this process—called the *jeitinho* in Brazil—has been developed by the anthropologist Roberto DaMatta (1991) and his student Lívia Barbosa (1994). The result is much less stability than the Japanese system. The concatenations of horizontal-vertical bonds that actors can assemble through their personal networks provide a counterbalancing force for the authority of the patron and can lead ultimately to the emergence of new patrons and factionalism.

Beyond the complicated negotiations of hierarchy and personalism in Brazil is another dynamic that gives the system its overall structure. According to the thesis developed by DaMatta in *Carnivals, Rogues, and Heroes*, the Latin American system is complicated because these traditional structures of hierarchy and personalism are crosscut by modernizing, Western institutions based on principles of equality and individualism. At the present time, then, Latin American society is like other societies where there is still a strong legacy of European colonialism, for example, India and some African countries. Drawing on the work of Dumont, DaMatta has described these societies as constituted by a unique dynamism of two clashing sets of structures: one modern, individualistic, and egalitarian and the other traditional, holistic, and hierarchical. As a result, there is a second mitigating factor for authoritarian hierarchy, one that suggests structures similar to those described for the United States.

Thus, Latin American culture is quite different from that of the United States but not at all in the same ways that Japan is. Is there any evidence that such fundamental differences in the social structure play themselves out in the social organization of Latin American scientific communities? In my own experience with academic anthropology departments in Brazil, I saw how the official side of collegial governance was contravened by factionalism. Although factionalism occurs in American departments, and Nakane describes it as a risk in Japanese organizations, I think it is probably more common in Latin America. Furthermore, from my experience the student/faculty relationships were much more hierarchical than in the United States; students were much more under the thumb of their faculty advisors, who were indispensable because they protected the students from other faculty members.

In Brazil when faculty lines open up, there is a search, so the process has the appearance of a modern competition as in American searches. However, the modernity of the competition is mitigated by a number of factors. First, there is less geographical mobility in general in Brazil, so searches tend to play out on less of a national scale. The reason why there is less geographical mobility is that people tend to have close regional ties, and they are less willing to leave their extended family networks, with which they are closely identified, in order to strike out on their own as individuals in another city. In addition, departments are less willing to hire people whom they do not know or whose advisors they do not know. Thus, the student is much more dependent on the advisor, and the advisor-student relationship has much more of a quality of a patron-client tie. Furthermore, once people are hired into a position, they are hired with "tenure." Thus, there is no six-year grace period in which junior faculty may move on to other positions once they have better established themselves. Instead, they tend to be locked into one institution, where they are likely to spend the rest of their academic lives.

Finally, there is little in the way of a reward structure that would benefit researchers who have, in the style of individualistic academic entrepreneurs, built up reputations through research and publishing. Indeed, by building up a reputation and publishing a great deal, Brazilian researchers may only excite the jealousy of their peers rather than earn a job at a better institution, which often offers the same salary structure as lower-status colleges. Because of the economic crisis, wages are so low that faculty often have to supplement their incomes through other activities, including teaching in other schools and universities. Personal ties as much as a publishing record govern one's ability to get teaching positions in other schools and in general to find supplementary employment to ensure an income that is marginally above the poverty level. Furthermore, the government is so broke that the grant game takes place with much lower stakes. In general, then, the apparently modern, individualistic, and egalitarian aspects of the Brazilian system tend to be weak, and one's security rests to a large extent on personal ties.

I found no suggestion of an exchange system in Brazil equivalent to the one that can be isolated in figure 5.2 for American anthropology departments. There was a sense that certain departments were forming an elite—Campinas (UNICAMP), São Paulo (USP), Rio de Janeiro (UFRJ), and Brasília (UNB)—similar to Harvard, Chicago, Berkeley, and other elite schools in the United States. Students with master's degrees from those universities often found jobs in schools of lower prestige, and often in the same region. However, although it was true that there were increasing numbers of Brazilians who were earning their Ph.D.'s at Brazilian universities, there was also a widespread sense that Ph.D.'s from a North American or European university had higher prestige. In discussions of foreign versus domestic graduate study, I noted the same complex negotiation of contradictory value systems when it came to the issue of domestic versus foreign Ph.D.'s. For example, I knew of one person who held a faculty position at an elite anthropology department where the senior faculty had earned their Ph.D.'s abroad. This person was held up as an example of breaking the chains of international dependency, and he was universally recognized as brilliant. Although the brilliance of the young professor seemed to prove that an indigenous Brazilian Ph.D. was as good as any other, at the same time his brilliance may have also been seen to compensate for the fact that he did not get his Ph.D. from one of the colonial centers. In general, my experience with Brazilian anthropologists is that even though there are now Ph.D. programs in Brazilian universities, they still prefer to go to the United States, France, or Britain. Presumably a foreign Ph.D. will give them additional prestige and career advantages when they return to Brazil. Given the long-term financial problems of that country, many hope they may never have to return.

A case study might provide a clearer sense of some of the contradictions of the social relations of science and technology in Brazil. The political scientist Antonio Botelho spent seven months doing an ethnographic study of a microelectronics laboratory in the state of São Paulo (1990). His research is one of the first studies to show how macrosociological processes such as the background of a military dictatorship and an ongoing economic crisis have affected the social organization of a laboratory. Although his account focuses on this very interesting phenomenon, it is also possible to read his ethnography from a cultural standpoint in terms of the conflict of individualistic, egalitarian structures with personalistic, hierarchical ones.

The conflict is evident in Botelho's story of how he got permission to work in the laboratory through a government-controlled Special Secretariat of Informatics (SEI). In Brazil microelectronics researchers aligned with some parties in the military government to make their research a national security issue that SEI oversees. SEI directed Botelho to work in a lab associated with a university in São Paulo, even though the agency also sponsored its own microelectronics laboratory. The government's laboratory was directed by someone who had split off from the São Paulo lab in an internecine quarrel, and thus the government agency may have decided to use Botelho to get more information about the rival São Paulo lab that the agency was in theory supporting. To avoid being considered a spy from the government, Botelho then activated personal ties that gave him a connection with someone who worked in the São Paulo laboratory.

Botelho's tale of entry is highly suggestive of what DaMatta has called the Brazilian dilemma, because Botelho in fact gained entry into the laboratory through two pathways: one through a universalistic process (he wrote a letter to a government agency and received permission) and the other through a personalistic one (a friend's brother). However, even within the universalistic, modern side of the picture, it is likely that the very fact that the government agency expressed any interest in him at all was due to rivalries between the two laboratories that were fueled by personal antagonisms. Furthermore, even though Botelho gained entrance to the lab in a relatively impersonal way, the entire process of getting permission took place in a national context of a military dictatorship (1964 to 1984) and a government policy that made microelectronics a national security issue.

Once he began to work in the laboratory, Botelho also found tensions in the style of different work groups. One group was focused around "a boss dictating what researchers should be doing, assigning tasks, or negotiating with finance agencies," whereas another group consisted of a younger generation of researchers with a more self-consciously democratic ideology that emphasized teamwork (1990:16, 20). The second group consisted of a younger generation because there had been a factional quarrel in which a large number of researchers left, and an older mentor figure later left Brazil to return to his home country. Thus, the relatively democratic, egalitarian condition of the second research group was itself a historical product of personalistic factionalism and international dependency relationships.

As for Brazil, Botelho's microelectronic research community seems to have only half-freed itself from the legacy of the *latifundia* and Iberian Catholicism. The factionalism and fiefdoms that he reports are similar to what I have seen or heard about in several Brazilian university departments. The operative principle is still very much the patron-client tie, here seen in the relationships between the patron and the employees as well as between the government-as-patron and the laboratory-as-client. Patrons dispense favors to their clients (personalism) and expect loyalty in their feuds with rival patrons (hierarchy). Nevertheless, there are always signs and suggestions of more democratic and modern arrangements—as, for example, among the second work group of the laboratory.

Similar problems characterized the Institute for Research in Biomedical Sciences of the National University in Mexico City, as described by the psychologist Jacqueline Fortes and the STS researcher Larissa Lomnitz (Fortes and Lomnitz 1994, Lomnitz 1979). They note that the organization had a very authoritarian structure, with power centered on a director who could fire department heads and allocate resources. The organization also suffered from intense factional quarrels at some points in its history. Usually those quarrels took place along departmental lines: some departments aligned with others in opposition to other factions. Furthermore, Lomnitz noted that collaboration tended to be hierarchical; that is, scientists tended to collaborate only as junior partners in a relationship headed by a senior partner. She noted that there was little horizontal collaboration—that is, voluntary relations of collaboration among people of similar rank—but she did find some cases of successful horizontal collaboration among younger scientists and scholars. As for Botelho, Fortes and Lomnitz found patterns of personalistic factionalism as well as signs of horizontal and cooperative relationships among younger researchers. The complexity of relationships and structures seems characteristic of Latin America, and perhaps for universities in other regions of the world outside the developed West.

Conclusions

The comparative social relations and social structures of technical and scientific communities could easily become the topic of an entire volume. Each country is in itself subject to huge variations across regions, genders, classes, ethnic groups, and disciplines. Although there is a fairly substantial STS literature of laboratory studies, those "ethnographic" studies have not examined in great detail the broader questions of the comparative social organization of research groups, departments, universities, and national research communities. As a result, more definitive statements on the topic await further research.

Rather than try to be comprehensive, I have chosen instead in this chapter to articulate some of the general principles that can be helpful in interpreting intercultural communication and comparative social structure. I have also provided a few case studies to give an example of how one might proceed from the general communication principles and national social structure to an analysis of differences in communication styles and social organization among scientific and technical communities. Methodologically, the background research should be used to provide a framework for getting a handle on concrete cases rather than a template or straitjacket into which the cases are forced.

6 | Science and Technology at Large: Cultural Reconstruction in the Broader Society

The remaining chapters explore knowledge and technology as they are understood outside the communities of technical experts. In this chapter I examine some ways in which the work, ideas, and technologies of scientists and engineers are understood and at times reshaped by interaction with nonexperts. This form of the cultural reconstruction of science and technology is somewhat more general than the form discussed so far, which has been limited mostly to reconstructions by experts within their own scientific and technical communities.

Questions of what constitutes valid, good, and successful science and technology are value judgments that frequently become very heated debates when scientists and engineers encounter other groups in society. Within this general field of debates one issue is what philosophers call the "demarcation problem": how to decide what is and is not science. One way to solve the problem is to refer to "reasonable" philosophical criteria that somehow can be used to distinguish or demarcate scientific inquiry and knowledge from other domains of thought. However, the strategy runs into a problem: who decides what constitute "reasonable" philosophical criteria? One answer is that reasonable philosophical criteria would be those acceptable to scientists, engineers, and other technical experts. However, there are many groups of people who do not accept the authority of the technical experts, and furthermore the

STS literature is full of examples that show how even the experts do not follow most of the standard reasonable philosophical criteria.

Cultural anthropologists who study technoscience have contributed to a different approach to the problem of demarcation. Rather than attempt to derive universal demarcation criteria through recourse to expert understandings of reason and reasonableness, we turn the problem into an empirical topic for research and ask how people go about making demarcations. Rooted in our methodological stance of cultural relativism, we assume that science and technology will have different meanings to different groups of people. Thus, we examine not only what constitutes acceptable knowledge or workable machines to the experts but also how the nonexperts understand, reinterpret, and pose alternatives to the same knowledge and machines. As technical discourses and devices pass out of their expert communities, they undergo a process of bricolage, and they are reconstructed according to the assumptions, values, interests, and politics of the other groups. Furthermore, managers, activists, religious groups, consumers, workers, and other nonexperts are often much more than relatively passive receptors who rework knowledge and artifacts produced by the experts. Instead, nonexperts are often actively engaged in producing their own knowledge and technologies, and thus reconstruction may involve active construction of new knowledges and technologies. The result is often a very different view of what counts as valid knowledge or successful technology. The two sections of this chapter consider some of the complexities of general reconstruction, first for science and then for technology.

The Cultural Reconstruction of Science

This section will focus on how science and scientists are reconstructed by nonexpert groups. Cultural anthropologists and cultural studies researchers have explored to date three major areas in which nonexpert groups reconstruct science and scientists. One area involves science and scientists as they pass out of the laboratory and research organizations into government bureaucracies and industrial enterprises. In the process scientists often find themselves and their science subservient to other interests, priorities, and ways of ordering the world. A second site involves cases of public controversy, in which scientists frequently find themselves painted as dupes of special interests who are forcing their projects on threatened communities. In these cases scientists are often pitted against not only community leaders but also other scientists who side with them. These situations provide a rich terrain for understanding how the reconstruction of scientific knowledge works. Finally, a third site involves religious groups, which sometimes develop alternative sciences, technologies, and especially medicines.

Science and Scientists in Nonacademic Organizations

Whether scientists work in universities and other large research institutions or in government bureaucracies and private corporations, there are institutional limits on their autonomy. Scientists usually work for someone else who is likely to be concerned with profit or organizational growth and survival. As scientists negotiate for their interests in organizations, they often find themselves demarcating boundaries between the scientist and the nonscientist manager or engineer. The nonscientist groups often have more organizational power or status, and thus the study of the boundary criteria in this setting could be seen as an example of reconstruction from above.

In the essay "Janus Organizations" the anthropologist Frank Dubinskas draws on his fieldwork in the United States in new biotechnology firms, where clashes between university scientists and entrepreneurial managers are especially pronounced. Among the principles he develops for such two-culture organizations, Dubinskas argues that culture clashes frequently crystallize over different assumptions about the nature of time. For example, scientists tend to think about their projects in terms of long-term research goals and benefits, whereas managers are more concerned with short-run, bottom-line notions of time. In addition to the confrontation over "planning time," the two groups frequently clash over what Dubinskas calls "developmental time." Scientists tend to see themselves as part of an ongoing process of development that leads through various ranks, honors, prizes, and publications to the eventual status of a senior guru figure for a whole field of inquiry. Managers, in contrast, tend to see themselves as having achieved a completed professional status and therefore as already being "developed." Whereas scientists view the managers as narrow-minded and myopic, managers view scientists as immature dreamers who are out-of-touch with the "adult" world of financial constraints and deadlines.

Dubinskas argues that the socialization process in graduate school shapes the dramatically different cultures. In the United States business school is only two years long, and it focuses on developing rapid analysis of problems that have fairly well-defined parameters. In contrast, graduate school in science involves four to five years or more, frequently with long days in a lab. Furthermore, in graduate school problems tend to be more open-ended. The doer/seeker difference between managers and scientists frequently plays itself out in young companies, where the two groups clash on opinions of how and where best to invest resources. Managers in general are the ones with the power, because they can hire and fire scientists. Thus, the doer/seeker boundary tends to favor the manager, who is constructed as practical, active, and adult in a culture that values those characteristics. However, managers are also dependent on the scientists to achieve their goals, and scientists may leave if the situation becomes unbearable. As a result, there is considerable room for maneuvering and negotiation on both sides. As Dubinskas shows, both managers and scientists are quite aware of the values, discourse, and rhetorical strategies of each other, and they often code-switch during negotiations. Thus, a scientist may justify a research strategy by speaking the language of investment and profits.

The anthropologist Stacia Zabusky (1992) describes a similar situation of clashing cultures in her ethnographic study of staff scientists in the European Space Agency (ESA). The situation is also somewhat different from Dubinskas's Janus organizations because the ESA is not a for-profit enterprise. Zabusky shows how scientists in organizations face the difficult job of negotiating not only with managers, engineers, and other pragmatists of the organization but also with their scientist colleagues who have positions in universities or other research organizations. Thus, the staff scientists of the ESA see themselves as part of the community of space scientists located in research organizations and universities throughout Europe, but at the same time they are part of the ESA bureaucracy. Located betwixt and between the two worlds, the staff scientists are put in the role of "ambassadors of science," translating between the needs and desires of the community of European space scientists and those of the ESA managers and engineers. The staff scientists experience tensions between their ideal of the scientist as someone who does research and the reality of much of their ESA work, which involves travel, paperwork, and meetings.

As in the biotechnology firms studied by Dubinskas, the culture of the intergovernmental agency is dominated by managers and engineers, and within the agency the managers and engineers view scientists as inferior. To them scientists are dreamers who are not able to make machines, solve problems, and approach the world pragmatically. To compensate, the staff scientists play up their abilities as pragmatists, and therefore as different from and superior to (within the framework of the managerial/engineering culture of ESA) their colleagues in the outside space science community. However, with respect to the outside community the staff scientists are likely to be perceived as inferior precisely because of their contamination with the pragmatic world of budgets and bureaucracy, and they run the risk of losing their status as members of the scientific community. Again to compensate, they talk about their role in terms of "service" to the scientific community (a relationship that assumes they are part of the community) and work on research projects with the community scientists.

A related study by the sociologist Stephen Barley (1988) focuses on conflicts between between X-ray technicians and radiologists in radiology departments of community hospitals. Barley invokes Hall's distinction (as discussed in the previous chapter) to describe X-ray technicians as monochronic relative to the more polychronic radiologists. The description also seems apt for the clashing cultures of managers and scientists as described by Dubinskas and Zabusky. In other words, the managers tend to be relatively monochronic—they have an orientation toward the schedule and doing one thing at a time—whereas the scientists have multiple projects and are less concerned with schedules and deadlines. Thus, the categories, concepts, and terms discussed in the previous chapter as differences across national cultures might also be employed to understand better the differences across professional cultures within organizations. At the same time, by comparing organizational and national cultures it is possible to see how, for example, the managers' relatively monochronic approach to time aligns them with the dominant pattern in the general culture.

The study of cultural boundaries between scientists and nonscientists in organizations has a greater significance, however, than showing how cross-cultural national differences dovetail with professional ones. In society at large scientists are often construed as elites and authority figures. Yet, in their organizational settings they are often placed on the defensive with respect to managers and other professional groups that are more closely allied with capital. Within this setting scientists may then find themselves in the position of an internal professional counterculture that expresses opposing viewpoints and values to professional groups that have their own uses and place for the scientists' technical expertise.

The situation described by Dubinskas and Zabusky for nonuniversity organizations is increasingly likely to apply to universities as well. In the United States the old vision of a university as a community of scholars and scientists organized along democratic principles is increasingly cast aside in the race to the technoversity, in which profits come from high-tech ventures such as biotechnology institutes. Education is still valued, if only as a tax-dodge to preserve a nonprofit status. As the business/university relationship tightens, TQE (total quality education), restructuring, reengineering, downsizing, and other new managerial fads and techniques become part of the administrative culture, and departments are forced to justify their existence and budgets in economic terms. Even the universities, then, have come to resemble Dubinskas's high-tech Janus organizations.

Science and Scientists in Public Controversies

The second context for the reconstruction of science moves in the opposite direction. Here, I consider public controversies in which government or corporate scientists are positioned as powerful or as representatives of the powerful in conflict with relatively less powerful representatives of the general public or threatened local communities. In this situation boundaries between the scientific and the extra- or nonscientific are constructed in tandem with boundaries between dominant groups and local communities.

Perhaps the most commonplace area of struggle over local resources involves environmental issues. Since the 1960s the environmental movement has undergone a rapid change as environmental organizing has become linked to local struggles for social justice. What was once a fairly romantic movement that sought controls on pollution and guarantees of natural preservation has become politicized through NIMBY (not-in-my-backyard) struggles over pollution and waste sites. As the historian Robert Bullard has argued in *Dumping in Dixie*, in the United States the historical structures of racism have often turned NIMBY into PIBBY—place-in-blacks'-backyard. Of course, PIBBY is not restricted to African Americans. As the growing and diverse movement for environmental justice demonstrates, environmental politics in the United States now operate explicitly in a context of Native American, African American, Asian American, and Latino/a organizations as well as class-based, gender-based, or neighborhood-based coalitions.

In many of the cases involving NIMBY and PIBBY controversies, local communities reconstruct official assessments of environmental safety and risk. In the process a variety of social differences become aligned with technoscientific ones. Those social differences include not only class, race, and locale but also some very complex and often contradictory social categories. For example, sometimes the health of children becomes pitted against the factory jobs of adult workers in ways that divide the community in its resistance to corporate constructions of safety and risk.

In addition, there is also increasing evidence that environmental injustice represents a double injustice to women, especially to women from underrepresented ethnic groups. Even traditional women's occupations—from nursing and office work to housework—are now well documented as sites for environmental hazards. It is not surprising, then, that many of the leaders of the movement for environmental justice have been women. For example, in the United States the Cherokee group Native Americans for a Clean Environment is led by women. In India women have also played a prominent role in movements for environmental protest.

In controversies involving local control over resources, communities frequently face the task of reconstructing apparently "objective" scientific accounts that they receive from government agencies or corporations. In the process, boundaries between what is and is not scientific are drawn against a backdrop of local and nonlocal interests. The anthropologist Gary Downey (1988) describes one example that involved a NIMBY controversy over a proposal to locate a nuclear waste site in the state of New Mexico. The Department of Energy (DOE) sponsored the project, and it enlisted support from scientists from the U.S. Geological Survey and the Sandia Laboratories, a federal research facility that DOE had asked to evaluate the site. DOE also drew on reports from the prestigious National Academy of Sciences. A number of local citizen action groups protested the site, and they found support from two university scientists, including an emeritus professor from Berkeley. The opposition groups also relied on scientific evidence and technical knowledge provided by a citizen action group, which in turn relied on

DOE documents. Downey argues that the strategy of using government data provided the opposition with credibility because their data could not be assailed. Instead, they focused on finding inconsistencies in the government position. By pointing toward inconsistencies in the government's scientific account of the proposal, and by using the research of the university scientists, the opposition built up an alternative scientific account. In this way the opposition transformed an official "scientific" account into a nonscientific one.

The anthropologist Priscilla Weeks (1992) found similar strategies in a controversy between government science and coastal fishers. The controversy emerged when the Texas Parks and Wildlife Department (TPWD) decided that oyster resources were near depletion, and consequently the department decided not to open the oyster season. In turn, the oyster industry filed and won a lawsuit to force the government to open the season. During the trial the oyster industry attacked the accuracy and interpretation of the TPWD's scientific data. Among the arguments against the TPWD, the oyster fishers and their supporters challenged the sampling methods and definition of depletion. They also noted that because oysters produce up to a million eggs a spawn, it would be possible to repopulate a reef easily as long as suitable shell is available. In other words, they argued that the government decision was based on flawed science.

As in the New Mexico case the local community recruited scientists to its cause. Unlike in the New Mexico case, the scientists on both sides of the Texas oyster dispute were part of a closely knit community. One of the two key scientists who testified against the TPWD decision was an emeritus university professor who had been the teacher of the chief of the TPWD Fisheries Division, and the other key scientist was a retired biologist who for thirty years had worked for the same chief. Thus, although the Texas fisheries case involved a dispute between fishers and the government, it had more of the flavor of a family quarrel than the New Mexico conflict.

Although there were some differences between the New Mexico and Texas cases, the parallels are also instructive. By recruiting scientists and learning the science, the two local communities reconstituted themselves from a position of nonexperts outside the game of science to a coalition of scientists and others with local knowledge expertise. In other words, part of the strategy to achieve political power involved translating their political position into a scientific language. The strategy is parallel to that of the scientists in the Janus organizations who translated a research objective into a language of investments and profits. In both the Janus organizations and the environmental groups, the less powerful group improved its position by using the scientific discourse of the more powerful group.

It is also suggestive that in both the New Mexico and Texas cases the threatened communities found support from university scientists who were not part of the government grant pipeline. Within this group they opted for emeritus or retired professors, people with less to lose by taking on the government but also with the status and prestige to do so credibly. In addition, the threatened community in both cases tended to build its alternative scientific account by focusing on contradictions and inaccuracies in the data of the government scientists. As Downey points out, that strategy enhanced their credibility by preventing their position from being undermined by attacks on the quality of their own data. Furthermore, the reconstruction of government science accompanied a reconstruction of government scientists: the oyster fishers suggested that the TPWD decision might have been influenced by connections with recreation and oil interests, and the opponents of the nuclear waste site in New Mexico painted the proposal in terms of big government taking control over their lives.

It is helpful to add one other case to this discussion, that of AIDS treatment activists. The case shows that the patterns just discussed are not restricted to the environmental movement but can extend to the activism of almost any sort of disempowered group in which science is a key ingredient in the conflict. AIDS treatment activists, as well as activists in the women's health movement, take the process of reconstruction one step further by becoming actively involved in changing the rules of scientific production. As the cultural studies researcher Paula Treichler shows in the essay "How to Have a Theory in an Epidemic: The Evolution of AIDS Treatment Activism," AIDS activists have gone far beyond recruiting scientists to their cause and challenging official accounts by pointing to contradictions within those accounts or the interests that shape them. Although AIDS activists use those strategies in their reconstruction of AIDS treatment research, they have also come up with new treatments, their own research on treatment efficacy, and their own methods for determining efficacy. In the process the activists have revealed the politics of the "objective" truth-finding methods of the prospective, blind, clinically controlled trial, and in their place they have developed more flexible and humane alternatives. As Treichler argues, their work points to the possibility of a radically democratic technoculture (98). The lessons of technoactivists of all sorts points to the possibility of a political culture in which experts are not the only ones who have the power to draw boundaries between the scientific and nonscientific.

Science and Scientists Among Religious Groups

A third scene of the cultural reconstruction of science and scientists involves alternative accounts that emerge from religious groups. The official religions of most Western societies seem to have achieved, after a long history of conflict, a more-or-less stable set of boundaries that allows scientists to be experts at science and clergy to be experts at the spiritual. However, followers of heterodox religious beliefs often do not honor those boundaries; indeed, much of their vitality—and controversial status—derives from their conflation of the boundary between the scientific and the nonscientific.

Generally, heterodox religions in Western societies fall into two groups on the issue of science: fundamentalist and occultist/spiritualist. The fundamentalist religions continue the tradition of opposition to scientific knowledge that characterized the rejection of heliocentrism and the persecution of Galileo. In general, those religions develop oppositional systems such as creation science or faith-based alternative medicine. I find occultist/spiritualist groups more interesting because, rather than oppose cosmopolitan science, they tend to draw on it and extend it in creative, if not always credible, ways. I will focus here on one example from these religions to give a sense of the nature of the cultural reconstruction of science by alternative religious groups.

It would seem reasonable to assume that occultist and spiritualist religions are on the decline, but most of the advanced capitalist societies seem to have become paradoxically more rationalized, scientific, and technological and at the same time more open to alternative religions, systems of healing, and techniques of spiritual development. Indeed, the juxtaposition of those apparently opposing beliefs is part of the postmodern condition. Although skeptics, rationalists, and many scientists—not to mention the orthodox clergy—have rejected alternative knowledges and technologies of this type as "pseudoscience," most of the surveys and polls indicate that their warnings have little effect on what the rest of the population believes. A cultural perspective would suggest that a better strategy than outright rejection and debunking would be to understand first the nature of the alternative knowledges that are being proposed, and then to assess whether or not there is any value in them. I will consider one case drawn from my own fieldwork among Spiritists in Brazil, who see themselves as antioccultist rationalists who have found a scientific basis for religion.

As I have argued in *The Brazilian Puzzle*, in some ways Brazil represents the future for the developed West. Not only is Brazil a multiracial and multicultural society, it is also a society of racial intermarriage and cultural blending. Brazilians—even many devout Catholics—honor the African *orixá* spirits as much as the Catholic saints. Furthermore, as many as ten million Brazilians (or 5–10 percent of the population) are Spiritists. They believe in communication with the dead via mediums (or "channelers"), the existence of a spiritual body that can govern the health of the physical body, and spiritual purification through reincarnation. Spiritist doctrine is based on a theistic, Christian moral system that was developed by a nineteenth-century French educator known as Allan Kardec. Spiritist intellectuals are interested in what they see as the scientific aspects of Spiritism, such as the study of psychic phenomena associated with mediumship. They have produced elaborate discussions of biology, physics, and the natural sciences. The result is a rather substantial body of publications that reconstruct Western sciences from a Spiritist perspective.

Probably the most respected version of Spiritist science is the work of Hernani Guimarães Andrade, an engineer who lives in São Paulo. His main theory begins by arguing that the body is capable of generating electric fields that cannot be explained by neural activity. He points to research on electric potential differentials at acupuncture points, Soviet claims about "bioplasma," and American studies of direct current electric fields. He then postulates a "biomagnetic field" that surrounds the body and would be capable of generating such electric fields. Andrade further argues that the biomagnetic field is generated by an invisible spiritual body that exists in a parallel space-time universe. Because the biomagnetic field exists in this parallel universe, it is not directly observable; it can only be observed on the physical plane through its effects, such as body electric fields. Andrade also argues that the biomagnetic field governs the patterns of growth and regeneration for organisms. He believes that the field can solve several claimed anomalies in evolutionary biology, such as the problem of homologous organs (different genes that result in similar features across species), orthogenesis (the tendency for evolutionary developments to follow a linear logic), the existence of instincts, and the patterns of interspecies mimesis.

Brazilian Spiritism has also become a fertile soil for the growth of new alternative therapies. In the 1980s Spiritists cosponsored international conferences on alternative medicine, and Spiritist mediums have long provided a variety of alternative healing services to the Brazilian public. As activists Spiritists have also been engaged in the struggle for the legalization of spiritually based alternative therapies, which the medical profession and Catholic church have opposed and the state has repressed. At various points in their history Spiritists' ability to translate their spiritual beliefs and healing practices into scientific theories has helped legitimate their struggle against repression.

For Andrade and other Spiritists their theories are self-conscious expressions of their identity as Brazilians who do not completely accept the materialistic outlook of the modern West and the orthodox medical profession in Brazil. Surrounded by African, Native American, and Iberian spiritual traditions, Andrade and other Spiritists urge the world to take a second look at the longstanding Western counterculture of vitalism and spiritualism known variously as spiritualist, occultist, metaphysical, and now New Age thought. Viewed as the trash heap of Western culture, these doctrines are the source of a unique vantage point on the dominant culture, one that takes on new meanings in the Brazilian context.

One may or may not agree with the Spiritists' reconstruction of science and medicine. What it *does* is remake science and medicine in a way that makes sense to people who live in a culture in which spirits are as much part of the reality as are televisions and trees. Like the environmental and AIDS activists, Spiritists disturb the boundary between science and nonscience by learning to talk science, even if they enter the world of cosmopolitan science through its heterodox backdoor of parapsychology or bioelectromagnetism. In the process Spiritists disturb the boundaries between first world "science" and third world "belief" that Westerners continue to draw in their discussions of rationality.

Of course, after having entered the Spiritists' world and situated it, one may still choose to reject some or all of their claims. However, in the case of the Spiritists I am skeptical of facile rejection via debunking. There are other ways to approach the Spiritists' intellectual project than either agreeing or disagreeing with it. For example, stepping out of the frame of orthodox science and into alternative frames such as that of the Spiritists provides a good vantage point for examining the deeper cultural assumptions that constitute the boundaries of orthodox cosmopolitan science. New types of questions, but reconstituted within the confines of what one defines as "acceptable" science, are possible. For example, one might do empirical research on biomagnetic fields or ponder more deeply the possible causes for order behind the apparently random mutations of evolutionary theory. Historically, groups that have been labeled occult have occasionally been the source of new ideas that eventually make their way back into orthodox science and medicine, albeit after the filtering and sanitizing process of a second wave of re-reconstruction by those scientists who pay attention.

The Cultural Reconstruction of Technology

Just as scientific knowledge and scientists are reconstructed by other groups, so technology is reconstructed as it passes from those who produce it to those who use it. This section will consider user reconstructions by two of the most salient groups: consumers and workers. Consumers have much more choice and control over new technologies than do workers, but at the same time, consumers are frequently more fragmented unless they coalesce into product-oriented consumer groups. As a result, the set of questions that emerges around consumer reconstructions is quite different from those that emerge around worker reconstructions. For workers the new technologies of the workplace are often handed to them without much consultation, and the technologies also frequently alter workplace relations.

Consumer Perspectives I: The Problem of User Acceptance

Artificial intelligence research has become the topic of a number of studies in the various constituent disciplines of science and technology studies. Anthropologists such as Lucy Suchman (1987, 1990) and Diana Forsythe (1990, 1992, 1993) have added to the debate by examining and providing a cultural critique of the AI research community's assumptions about the nature of intelligence. As the products of the AI researchers move from laboratories to workplaces, the narrow assumptions of the AI programmers clash with the highly context-dependent, general assumptions of their consumers and customers.

In the essay "Blaming the User in Medical Informatics: The Cultural Nature of Scientific Practice" Forsythe examines some of the unchallenged assumptions about knowledge acquisition in the area of AI research known as expert systems. Expert systems are computer programs that attempt to duplicate the knowledge of experts and to help technical and professional people to improve their decision making by comparing their decisions to the recommendations of the computer program. Forsythe examines the problem of "user acceptance" in medical expert systems (such as programs that aid in diagnosis), and she asks why expert systems are not widely adopted even though they have been in existence for more than twenty years and have received millions of dollars of government support for research and development. Forsythe finds unconvincing the "native" explanation of the AI researchers themselves. Their term for the problem is suggestive of their approach: they speak of "end-user failure" and think of the problem in terms of a public that suffers from computer phobia. Forsythe argues instead that expert systems tend to go unused because built into their programs are the naive assumptions that their producers have about how to acquire knowledge, what counts as knowledge, and how it is used.

In general, the programmers tend to value "hard" knowledge (formal, technical, and quantitative) over "soft" knowledge (informal, nontechnical, and qualitative), even though, at least for the consumers of expert systems, in real-life situations most useful knowledge is "soft." Forsythe argues that the programmers tend to think in decontextualized terms; to borrow the phrase of sociologist Susan Leigh Star, they "delete the social" (1991). As a result, the programmers tend to build their view of knowledge into their systems at the expense of more contextualized and socially laden knowledge. In building AI models of the experts' knowledge, the programmers tend to assume one reality—that there is one right answer to a problem—and that conscious models can be taken as accurate representations of the world. Finally, the programmers evaluate their systems not in terms of how they are in fact used but in terms of technical factors such as the speed or architecture of the program.

Forsythe's work suggests one practical application of the study of the cultural reconstruction of technology. The question is deeper than a simple problem of producers not designing products that match consumer needs. Rather, producers have a model about what knowledge is, and their model is a restricted one that is itself a product of their disciplinary culture. Forsythe suggests that a prerequisite to good programming, at least in this case, may be good anthropology.

Consumer Perspectives II: Women's Reproductive Technologies

Another area of consumer reconstructions of technology in which major research has now been done is women's reproductive technologies. As in the case of expert systems, anthropologists have contributed to interdisciplinary discussions in the field by developing a cultural perspective that analyzes questions of power. Of the many case studies in the field, I will select only three, those by the anthropologists Rayna Rapp, Robbie Davis-Floyd, and Linda Layne.

In several essays Rapp focuses on amniocentesis, a procedure that assesses potential genetic disabilities of the fetus. Her work is distinctive in its comprehensiveness: she points to a wide variety of groups in New York City that have different relations with the technology. The groups range from the experts, such as laboratory technicians and genetic counselors, to parents in support groups, differently abled rights groups, and women who go to counselors, have a positive diagnosis, or refuse the new technology. Rapp shows how the technology opens up different choices, opportunities, and dilemmas for all these groups. She then explores how the constructions of the technology shift dramatically for the users according to gender, age, class, education, religion, and type of genetic disease. Rapp notices how notions of risk vary widely, but also how user educational level varies and how genetic counselors shift codes according to the user's ability to work with sophisticated statistical ideas.

The anthropologist Robbie Davis-Floyd (1992a, 1992b) has also shown different uses and understandings in her study of high-tech hospital birth technologies. In a study of middle-class women in Texas, Davis-Floyd shows that while some women opt for natural birthing and midwives, others opt for high-tech hospital birth. Her research reveals that high-tech hospital birth is not automatically disempowering to women; rather, the power effect depends on the cultural assumptions and preferences that women bring with them to the birthing process. In some cases professional women who are at home with the scientific discourse and technologies of the hospital feel they have more control over their bodies in the high-tech hospital setting than if they surrender to natural processes and midwives. The sense of empowerment and birth choices depends in large part on the way the women think about their bodies. Through interviews Davis-Floyd shows the home-birthers think of the self and body in continuous terms, whereas the hospital birthers are more Cartesian. The central point is a similar cautionary tale about universal models of new technologies. Like Rapp, Davis-Floyd shows how the same technology or set of technical procedures often has different meanings to different groups. Thus, the key to the question of evaluating new reproductive technologies and their impact on women's power may have less to do with introducing or regulating the technology and more to do with preserving the woman's right to choice and control over her own body. Furthermore, as Rapp shows, once the choices are available, women and other consumers of new technologies need to be able to choose intelligently among the options, and that in turn depends on access to counseling and education.

The anthropologist Linda Layne (1992) discusses the ways in which another reproductive technology, ultrasonic imaging, has posed new problems and dilemmas for expecting mothers and fathers. She shows how the technology comes to mediate the mother's experience of her pregnancy and can sometimes lead to what Layne calls "dis-synchronicities" between the mother's experience and the knowledge she has of the fetus. For example, the technology can inform the mother that the fetus is dead before she experiences a miscarriage or the stopping of movement. Furthermore, because the parents are able literally to see the fetus, and even to take home an ultrasonic photograph of the fetus, they tend to establish an earlier and closer identification with the fetus than existed before the technology came into use. The result is that pregnancy loss tends to be experienced in similar ways to the loss of a baby, and parents' grief may be greater than it was before the imaging technology became available. The technological image of the fetus/baby also becomes intertwined with religious images of the baby as angel. Layne draws on fieldwork among pregnancy loss support groups and on their poetry and writings to show how parents reconstruct the medical discourse and ultrasound technology by switching back and forth between technological and religious images of their fetus/baby. The result is that parents come up with an odd, if reasonable, construction of pregnancy loss in which ultrasound fetus images are intertwined with religious images of baby angels.

These studies all open up a cultural analysis of the new reproductive technologies from the point of view of women in particular and users in general. As Rapp writes, "Until we locate and listen to the discourse of those women who encounter and interpret a new reproductive technology in their own lives, we cannot evaluate it beyond the medical model" (1991:392). Feminists have long taught that the personal is the political; these newer feminist studies show that the technical is also the political.

Worker Perspectives I: New Managerial Technologies

In *Labor and Monopoly Capital* the political economist Harry Braverman discusses the ways in which work has become increasingly subjected to scientific management principles. Since the Taylorist project of scientific management at the beginning of the century, managers and engineers have worked together to chart, time, map, and otherwise rationalize the worker's labor. The project makes it easier for management to replace workers and to supervise their efficiency, thus further enhancing the domination of management over workers. In turn, workers have developed interesting strategies of resistance. For example, the historian and culture critic Michel de Certeau (1984) describes a practice known as *la perruque*, in which workers surreptitiously use company resources and clock time for their own purposes, such as letter writing, phone calls, and so on. In third world settings resistance may invoke local cultural forms, such as factory women in Malaysia who become possessed by spirits, or miners in the Andes who construct a devil in the form of the white, male owners of the mines.

In the emerging global economy the model of assembly-line production invoked by Taylorism is increasingly outdated. In its place industrial companies are moving toward "flexible-systems" production in which they can quickly retool to adapt to rapidly changing consumer tastes. Workers are, in theory at least, no longer mere automata modeled on the Model T Ford with its replaceable parts. Rather, workers are part of the human capital of a company, part of its competitive edge, and a potential source of creativity. Buzzwords have switched from "time-motion studies" to terms such as "total quality management" (TQM), an ambiguous term that somehow links competitive success to product quality, product quality to worker input and involvement, worker input and involvement to the worker's sense of belonging to the company, and all of it to

Japanese managerial techniques. Under TQM scientists and engineers are enlisted in the struggle for improved quality in a production system that emphasizes constant technological and managerial innovation.

The new technologies of management and production are situated in a reconfigured world in which the United States, which seems always to have to have an ominous Other, now looks with fear and admiration to Japan. Because Japan is, more than any other country, associated with the new technologies of total quality management and flexible systems production, I will examine a case of Japanese management and managerial technologies in the United States. The overlay of Japanese/American and management/worker conflicts will help dramatize how the new "postmodern" style of technomanagement is itself interwoven with international cultural politics and political economy. Notice, in this case, that the expert culture is the management culture, rather than scientists per se, and the nonexperts are the workers.

By 1990 the Japanese owned or partially owned approximately 640 plants in the United States, and they employed about 160,000 people. Estimates have claimed that as many as one million Americans may be working for the Japanese or Japanese-owned companies by the year 2000. In the automotive industry, which is perhaps the best-known case of Japanese manufacturing in the United States, all the major Japanese companies have at least one plant operating in the United States. On the whole the Japanese have preferred to set up factories in the lower Midwest or the upper South, such as Honda in Ohio, Nissan in Tennessee, Toyota in Kentucky, and Subaru in Indiana. The Japanese prefer the small, rural communities of the heartland because they believe they can still find the traditional work ethic, and the states in the region have the additional benefit of being centrally located with respect to major population centers (Gelsanliter 1990). However, some critics have also charged that the Japanese have chosen those communities because they want nonunion employees and ethnic groups of predominantly Northern European descent. In any case, the "fit" between the paternalistic Japanese management style and the conservative nature of their host communities seems to have worked out well in most cases, and in at least two major votes the workers have even decided not to unionize their plants.

Nevertheless, in one case a Japanese company chose to locate in a heavily union area of Michigan, the home of American automobile manufacturing. In that case the relations between the Japanese management and the more militant and experienced American workforce soured substantially. The case of Mazda's factory in Flat Rock, Michigan, is therefore instructive, for even if it is the exception to the rule, the conflicts that occurred there reveal some of the connections between cultural differences and the new managerial technologies of TQM. As the journalists Joseph and Suzy Fucini describe in their book *Working for the Japanese*, Mazda apparently did not want to locate in a union area, nor did it wish to open a union shop, but it went union because it was 25 percent owned by Ford. The United Automotive Workers (UAW), however, was happy to have the plant open in the economically depressed region, and as a result it promised to be cooperative.

At first the technical and managerial innovations of the new plant seemed fairly innocuous to the workers. For example, the Japanese used a "just-in-time" inventory system, so that extra inventory and idle working time were kept to a minimum. The system is quite different from the conventional American style, which is sometimes referred to as "just-in-case." In other words, the American system keeps inventories high and builds in some idle working time just in case suppliers are late or the workers become injured or get sick (Fucini and Fucini 1990:36). A second innovation was the system of *kaizen*, or constant improvements, in which workers meet and make suggestions for improving the production process. *Kaizen* therefore involved worker input and, as the managers promised, consensus and team management. Thus, the system appeared to be more democratic than American production, and the workers were initially very pleased at the apparently more democratic system of the Japanese. The democratic rhetoric was bolstered by a number of egalitarian symbols: the top manager and lowest worker wore the same uniforms; everyone ate together in the same cafeteria; administrative offices were located in the middle of the shop area, which increased accessibility; and even the manager's desks were all located in the same room with low dividers between them.

However, problems emerged as Mazda began to try to increase its productivity. Injuries began to occur as the Japanese managers opted for increased efficiency over safety. The workers began to complain of back problems and trauma injuries such as carpal tunnel syndrome, an inflammation of the tendons in the hands and wrists that is caused by excessive repetitive motions. The workers discovered that *kaizen* suggestions for increased safety often required studies, whereas those that increased productivity were quickly implemented. Peer pressure grew quickly not to make *kaizen* suggestions because they only increased the team's workload. A worker who later became the leader of the dissident union faction summarized *kaizen* and the team system in the following terms:

> They were going to *kaizen* out this and *kaizen* out that, so we could be more productive. The more they talked, the more it sounded like this whole team thing was just a way to squeeze more work out of every worker, with a good dose of old-fashioned paternalism thrown in to keep everybody happy. (87)

The team system soon fell apart. Because team leaders were union members, the other members of the team tended not to follow the team leader's orders. As a result, the team leader of six to ten workers had to call on the "unit leader," a management representative who supervised several teams. Polarization between management and the workers became even more pronounced: "The bigger meetings were less personal, they provided each individual worker with less of an opportunity to contribute to discussions, and, of course, they were now under the direct control of the management" (133). In turn, the Japanese manager complained about the Americans' individualism, which he thought made it difficult for them to work as a team. Ultimately the highest-ranking American manager quit in frustration because the real decisions were all being made back in Japan.

Conflicts erupted over a number of other issues not directly related to the new management techniques, such as the sick leave policy and the poor record of the Japanese managers for hiring women and members of underrepresented groups, especially African Americans. In the end a group of dissident union leaders stood for election and unseated the pro-management union leaders. After coming to power the dissidents inaugurated changes in attendance policy and in the status of temporary workers, and the American workers and Japanese managers were able to work out some of their differences.

The case of the Mazda factory is, to repeat, an exception. In many factories American workers and Japanese managers have gotten along well, and the new managerial techniques seem to be relatively successful. However, in the case of the Mazda factory the meeting of the militantly individualistic (and patriotic) Michigan working-class union culture with a comparatively hierarchical and paternalistic Japanese management culture proved an explosive combination. To some extent the conflict can be attributed to clashing national cultures; in other words, the

Japanese model of the family (the *ie*), and the loyalty and duty that its members owe to it and its reigning patriarch, informs the Japanese management style. In contrast, Americans operate with their own cultural models of the nuclear family, representative democracy, and the social contract among equals. Issues such as sick days, the hiring of women and underrepresented groups, dress codes, health benefits, and the trade-off between safety and efficiency became arenas in which the rights of the individual and the duty to the whole, and with them two different notions of society, were negotiated as well.

At the same time, however, the case of the Mazda factory makes it possible to ask some questions of the new Taylorism of the late twentieth century. In place of time-motion studies managers in the 1980s and early 1990s were talking about total quality management, just-in-time inventory systems, and *kaizen* in the production process. An optimistic interpretation of the changes would be that management is finally beginning to see workers as human beings, as thinking creatures who can participate in and improve the production process. However, a more pessimistic view is that the old Taylorism, trapped in the behaviorism of modernist science, merely disciplined the bodies of the workers; the new Taylorism, imbued with the insights of cultural anthropology, social psychology, and other human sciences, is now claiming the hearts and minds of its workers as objects of its disciplinary power.

Worker Perspectives II: Computerization

Purists might argue that the new Taylorism represents managerial techniques rather than technologies, and in the Mazda case the material technologies of the assembly line were left in the background. An example of machine technology and worker construction that is receiving increasing attention is computerization. In *Computing Myths, Class Realities* the anthropologist David Hakken and the education administrator Barbara Andrews analyze computerization in Sheffield, England, during the mid-1980s. The case is interesting because Hakken had done fieldwork there during the 1970s, so the authors were able to examine the region longitudinally, much as some researchers have examined areas before and after electrification. Also, Sheffield had a militant working-class culture similar to that of Flint, Michigan, except that in this northern English region the working-class was historically tied to mining and steelmaking rather than automobile manufacturing.

Hakken and Andrews reject technological determinist perspectives, whether of the positive, "computopian" type (computers will lead to a better workplace and society) or the negative, "computropian" type (computers will make everything worse). In contrast to those versions of the computer revolution story, they propose that the relationship between computerization and social change should be a topic of empirical research. Their research reveals that the relationship varies by economic sector and cultural domain. Computerization may be the cause, the effect, or merely the correlate of social change. Furthermore, they show that in predominantly private sector activities (such as small retail businesses) computerization fitted into existing patterns of social change, usually accelerating change only moderately. In contrast, in state-mediated primary production such as the mining industry, computerization became part of a major social transformation. Likewise, regarding changing relations between women and men, they show how computerization was sometimes beneficial and sometimes harmful to women in the workplace.

The issue of worker reconstruction of computers in Sheffield can be approached through a comparison of two cases. One, the transformation of the mining industry, involves worker constructions that in some ways are parallel to those of the Mazda employees. In other ways perhaps a closer parallel is the battle between the air traffic controllers and President Ronald Reagan during the 1980s. In Britain the miners' union was arguably the most militant in the trade union movement, and thus the miners became a target for the anti-Labour government of Prime Minister Margaret Thatcher. As Hakken and Andrews argue, after the Conservatives came to power in 1979 they prepared for a major confrontation with the miners by building up coal stocks, converting a number of power stations to oil or gas, and switching coal transportation from British rail to roads. Then, the government rapidly instituted computerized mining technology, which was set up to centralize information in the hands of management. Finally, the government announced a program of rapid pit closures, thereby provoking perhaps the most bitter strike in Britain since World War II.

In the 1960s labor and the mining management had negotiated over computerized mining technology, which management ultimately abandoned because it turned out to be impractical. The new government's decision to reinstitute essentially the same technology during the 1980s was therefore made for political reasons, and the miners' union saw through the decision and chose to resist both the new technology and the overall union-busting program of the government. During the strike the state was able to use new computerized surveillance technologies to further weaken the miners' position. Unlike the Michigan workers but like the American traffic controllers, the Sheffield workers lost the strike.

The cases of Mazda automotive workers and Sheffield miners provide examples of worker reconstructions of management-introduced technologies that threatened established levels of employment or job satisfaction. As a result resistance struggles emerged that provide an example of local/global dynamics similar to the cases of the threatened communities discussed at the beginning of this chapter. I will return to the question of resistance again in chapter 8 in the context of development and indigenous peoples. However, at this point it is worth introducing the concept of resistance and reconstruction in a complex way that shows how reconstruction can be proactive as well as reactive. Another case of worker reconstruction of computerization discussed by Hakken and Andrews provides an example of a more proactive approach.

Hakken and Andrews introduce the concept of "culture-centered computing" as a way of thinking about proactive forms of worker reconstruction of computerization. For example, they discuss the Traffic Systems Cooperative, a business that does traffic equipment service and is supported in part by the pro-Labour local government. The cooperative set up its office computer system to do filing, accounting, and word processing but not design or worker monitoring. In other words, computers are not used to plan work flows or to *kaizen* the workers by ratcheting up productivity against their will. Rather, the system is used to cut down on administrative work and to allow its workers to spend more time on the road doing repairs. As one worker put it, "I'm in favor a giving me mates more time to walk in the sun" (173).

The process of computerization therefore varies radically depending on the social configuration in which it is embedded. The case of the Traffic Systems Cooperative is an example of computerization that is tailored to a relatively democratic workplace in which workers direct the uses to which computers will and will not be put. Of course, the success of the project is conditioned by the support of a progressive local government and a business structure of worker ownership and participation in management decisions. For most of us those ideal conditions are a long way from the current reality.

Conclusions

In this chapter I have developed a critique of what might be called a technocentric view of science and technology. Like ethnocentrism, a technocentric view conceives of science and technology only from the perspective of the expert communities that produce it. However, science and technology are in fact many things to many people. Experts are not the only ones who have understandings of science and technology; scientific ideas and technological products often have very different meanings to nonexperts. Even experts with a slightly different area of expertise (such as managers, engineers, or merely scientists from other fields) are likely to have radically different viewpoints. Thus, again I show a way of moving beyond the idea that science and technology are socially constructed—that is, shaped by and imbued with the social circumstances surrounding their production. Science and technology are also socially or culturally reconstructed by those people who use and remake technoscience: by managers, local communities, religious groups, consumers, workers, and so on.

7 | Other Ways of Knowing and Doing: The Ethnoknowledges and Non-Western Medicines

Why study the knowledge systems other than those of modern, cosmopolitan science and technology? One type of answer looks at the topic from a theoretical angle: by studying the knowledge systems of other cultures, it becomes easier to see the possibilities and limitations of the ways in which the human mind organizes the world. Other knowledge systems may provide new ways of thinking about old scientific problems, or they may raise new problems and suggest new methods and topics of inquiry. A second type of answer is more pragmatic: understanding the knowledge systems of other cultures may help the members of those cultures to resist colonization, either in the ideological or in the political and economic domain. For example, by understanding the complexity of local and indigenous agricultural and botanical knowledges, it becomes easier to resist the ideologies of prodevelopment groups (such as third world states and missionaries) that wish to "civilize" non-Western peoples. Furthermore, by studying non-Western natural knowledge it may be possible to build a better case against development projects that would destroy irreplaceable natural resources such as tropical forests. This chapter will focus on two forms of non-Western knowledges and technologies: those involving knowledge about nature in general and those involving non-Western medicine.

The Ethnoknowledges

Other knowledge systems can be divided into roughly two groups: those of literate, Old World societies and those of New World societies and indigenous Old World societies. The first group covers the sciences of the ancient and medieval Old World empires: Babylonian, Greek, Roman, Egyptian, Moslem, Hindu, Chinese, and so on. As I have argued, a growing body of scholarship suggests that for millennia the Old World buzzed with transmissions that followed routes of trade and conquest. Because there was so much interchange among the Old World societies, it is never quite clear whether they should be thought of as variants of a single system or as dialoguing local knowledges (in effect, those perspectives are two sides of the same coin). For those reasons, anthropologists who are looking for the greatest cultural distance from modern science and technology have often been more intrigued by the knowledge systems of the New World societies or the remote, indigenous peoples of the Old World.

There are many ways to define the terms *native* or *indigenous* peoples, but the definition of the World Council of Indigenous Peoples incorporates two basic ideas: they are prior occupants of the land and they have a distinct language and culture (Burger 1987:8). The concept of indigenous or native is therefore a relative one; Native peoples occupied the land prior to other populations that now share or claim their former lands, and their language and culture are relatively distinct from the other populations. (I shall use the term *Native* with a capital letter to put the peoples on the same level with Europeans, Asians, and so on.) The other ethnic group or groups usually comprise national or cosmopolitan society, which itself is frequently divided into dominant groups (such as Americans of European descent) and nonindigenous but nonmajority ethnic groups (such as American descendents of African slaves). With a few exceptions (such as Greenland, where indigenous peoples have their own nation-state), the dominant ethnic groups of the national society control the state, usually with some sharing of power with the other ethnic groups. In contrast, indigenous peoples have a relatively small voice in the decisions made by the nation-state. In countries throughout the world Native peoples frequently occupy the bottom of the social pyramid, with even less access to resources and political power than nonindigenous ethnic minorities.

The term *ethnoscience* is sometimes used to described the knowledge systems of indigenous peoples. The term is problematic for a number of reasons. Historically, anthropologists have used the word *ethnoscience* to describe a method for analyzing non-Western and traditional systems of classification and cognition. However, the word can also be used in the way I will use here: *ethnoscience* is an umbrella term to cover the study of various non-Western systems of knowledge, such as ethnoastronomy, ethnomathematics, ethnobiology (including ethnobotany, ethnozoology, and ethnopharmacology), ethnopsychiatry, and ethnomedicine. Another problem with the term *ethnoscience* is that all knowledge systems are culturally rooted, and hence even international physics might be described as a Western ethnophysics. Finally, to apply terms such as *science, astronomy, mathematics,* and *biology* to non-Western systems of knowledge means using Western categories to cut up knowledge systems in ways that may violate the manner in which those knowledge systems divide up the world. Given all the problems, I prefer to speak about "other ways of knowing and doing," and I use the term *ethnoknowledges* as a somewhat less problematic and more convenient heuristic umbrella term.

Theoretical Issues in the Ethnoknowledges

Probably the most well-developed of the fields of non-Western systems of knowledge and technology is ethnobiology. The field now has a professional society and several journals, including the *Journal of Ethnobiology, Journal of Ethnopharmacology,* and *Ethnobotany*. A major topic of concern has been non-Western classification systems for plants and animals. One of the most well-known studies is "Why is the Cassowary Not a Bird?" in which the anthropologist Ralph Bulmer studied the animal classification system of the Karam of Papua New Guinea. The cassowary is a tall, flightless "bird" similar to the ostrich. In the Western, Linnaean system of classification, mammals may have wings (such as bats), but they must suckle their young, have hair, or have warm blood. Consequently, the Linnaean system places the cassowary in the category of birds. However, the Karam classify animals according to whether they fly, live on land, or swim in the sea. Because the cassowary does none of these, it is put in an anomalous category.

It would be reasonable to expect that the study of indigenous animal and plant classification systems might lead to the collapse of the Linnaean system; however, to some extent the result has been the opposite. The Western system has largely expanded by gobbling up non-Western local knowledge and incorporating unknown species from ethnobiologies into its cosmopolitan system of categories. Furthermore, studies of a large number of non-Western taxonomies suggest that diverse cultures have developed taxonomies that operate according to procedures that are very similar if not universal. That research finding led to a question: are the similarities among the classification systems the result of an inherently structured natural world or similarities in the way the human mind organizes the world? The question is a variant of an old philosophical debate about the nature of human knowledge, and the answer of anthropologists seems to be similar to that of the philosopher Immanual Kant: both.

A related field of inquiry has examined similarities in human classification of phenomena of the physical world. For example, the anthropologists Brent Berlin and Paul Kay have demonstrated that color categories in languages throughout the world tend to fall into a general organizing pattern, much as do biological taxonomies. For societies that recognize only two categories, the colors are generally bright (white) and dark (black). In three-color languages the colors are always red, bright, and dark, and in four-color languages the additional color is either yellow or green. This pattern corresponds to some extent to the system of opponent processing in the brain, in which the visual cortex processes colors according to a bright/dark, red/green, and yellow/blue system. Of course, all human beings can see a wide range of color variations, but people need to have a vocabulary in order to recognize and conceptualize the variations.

Studies of non-Western classification schemes are interesting not only because they show similarities in the way the human mind constructs the natural world across cultures but also because they show possibilities for new ways of seeing the world. Perhaps the most widely cited example is the claim that the Eskimos have over a hundred different words for *snow*. Although it is true that some cultures have a more elaborate vocabulary than others for phenomena that are of great interest to them, the claim about Eskimos and *snow* is highly exaggerated. The discussion of Eskimo words for *snow* began with a brief mention by the anthropologist Franz Boas back in 1911, who only claimed that the Eskimos had four types of words. The discussion "snowballed" until textbooks in the 1950s were claiming that the Eskimos had many words for *snow*. By the 1970s references in the popular literature and the media had grown to claims of over one hundred, and a television weather report even claimed that there were over two hundred Eskimo words for *snow*. In fact, there are only two distinct roots for *snow* in any Eskimo language, and those words are roughly equivalent to the distinction between *flake* and *snow* in English.

Although the case of Eskimo words for *snow* is folkloric, the general point should not be missed. Groups of people and societies tend to pay more attention to aspects of their world that are important to them. Those aspects can be social, natural, or philosophical. When people pay more attention to something, they tend to develop a more refined vocabulary, make finer distinctions, and produce more knowledge. Thus, a better example for the case of snow than Eskimos might be skiers, who distinguish carefully whether snow is manufactured or natural, wet or powdery, recent or old, with or without a base, and so on.

Another example of different ways of seeing the world comes from the field of ethnoastronomy. In *Empires of Time* the astronomer Anthony Aveni describes how Western notions of time became increasingly concerned with precision, order, rigidity, and linearity. In many non-Western societies timekeeping focuses on seasonal cycles and is closely tied to agricultural or hunting and gathering activities. In the larger societies that had states and a priesthood, calendars and astronomical observations were often quite sophisticated. Perhaps the best example is the ancient Mayas, who developed a complex double calendar of an interlocking 260-day and 365-day cycle. The 260-day calendar (the *tzolkin*) consisted of 13 "months" of 20 days each, and it corresponded to the human gestation period, eclipses, and the appearances of Venus. In general, examining non-Western systems for ordering time can provide a good comparative basis for understanding deeper cultural assumptions about time in modern, Western societies. Rather than take for granted that, for example, time is linear, unidirectional, and progressive, it becomes possible to play with other ways of organizing time, including the cyclical patterns of many agrarian societies.

Pragmatic Issues

Although the study of ethnoknowledges can be useful for the problem of understanding human cognition or for posing alternatives to Western systems for classifying the natural world, it is by no means limited to those theoretical interests. There are also a number of practical or policy-oriented reasons for studying non-Western knowledge systems. One type of practical reason has to do with education. As awareness of ethno-knowledges increases, it may be possible to redesign educational programs that are better adapted to local cultural settings.

Those implications are probably most developed in the area of ethnomathematics, which in recent years has demonstrated the formal properties of indigenous practices as well as the local mathematical knowledge of nonmathematicians in national societies. Sand graphs, kinship relationships, games, puzzles, riddles, and abstract design all show a high level of formal reasoning and can even be modeled formally. Likewise, diverse groups such as U.S. urban shoppers, Brazilian children in slums, and South African carpenters have been shown to have their own mathematical strategies.

Understanding how mathematics is embedded in everyday life has enabled researchers to develop teaching programs that are more sensitive to the cultural backgrounds of the students. For example, the philosopher Helen Watson-Verran has developed a way of teaching mathematics to Australian Aborigine children that draws on the implicit mathematical knowledge they already have, such as knowledge of their complex kinship system. In Brazil some educational reforms have linked ethnomathematical teaching strategies to the alternative pedagogy of Paulo Freire, who advocated teaching in a way that involved actively using the knowledge of students and thus empowering them with the tools for drawing on their own informal knowledge.

The study of the various ethnoknowledges can also be linked to local struggles for social justice and environmental protection. For example, ethnobiology includes the study of plant uses in addition to the study of indigenous plant and animal classification systems. The field of ethnopharmacology examines indigenous plant usage and also makes chemical analyses of the pharmacological properties of the plants and their potential impact on human health. As a result of ethnopharmacological studies, particularly those of tropical forest plants, new drugs have been developed. Ethnopharmacology has the advantage of showing third world states and development interests that local flora and fauna have much greater long-term economic potential if left alone than if they are destroyed for development purposes.

The downside of ethnopharmacology is that indigenous knowledge of plants is not patented. The huge multinational pharmaceutical companies frequently draw on Native knowledge of the medicinal values of plants and animals in what the anthropologist Jason Clay has labeled the "last great resource rush" (1990). He gives an example that succinctly summarizes the problem: missionaries in Western Brazil passed on a sample of arrow poison to some botanists, who in turn passed it on to a U.S.-based chemical company that supported their research, and in turn the chemical company took out a patent on the poison's muscle relaxant properties. The indigenous group did not receive a penny of royalties, and the government meanwhile seized half their land.

The issue of intellectual property rights can mean the difference between extinction and survival for some Native peoples. The stakes are far from minuscule, for even royalties of a fraction of one percent can run into the millions of dollars in the large international pharmaceutical industry. With funding available at that level, many Native communities would be in a much better position to resist incursion from the outside. A number of indigenous and international organizations have worked to set up guidelines to protect Native peoples from exploitation, and indigenous peoples are also working to set up guidelines and protective measures. For example, the Kuna of Panama have developed a twenty-six-page manual with guidelines for scientific research in their area. Researchers work in cooperation with the Kuna and the Smithsonian Tropical Research Institute in Panama City, which acts as a clearing-house that also disciplines researchers who violate the guidelines. The Kuna provide one example of an indigenous people who have taken steps to protect their intellectual property rights. If those rights are honored by third world states and multinational corporations, it is possible that ethnopharmacological knowledge may be transformed into an economic resource that empowers indigenous peoples and simultaneously protects their lands.

A second practical aspect of ethnobiology is indigenous knowledge of agriculture. In many tropical forest areas indigenous peoples practice slash-and-burn horticulture. They clear small patches of the forest by burning, and they plant crops on the soil that is rich from the ashes. In general, tropical soils are very poor, and as a result the crops usually do not grow well after a few harvests. The solution for many forest peoples is to move on to another plot of land and start the process over again. The practice does little damage to the forest, because the first plot of land generally grows back fairly quickly.

Indigenous tropical agriculture usually involves a mixture of crops. In a one-acre plot of land, for example, there may be a mixture of two dozen or more plants ranging from tubers and beans to fruit-bearing trees that also provide some shade for the other plants. The mixture of plants has many agricultural advantages, but in tropical areas probably the most important advantage is that diseases that attack one plant often cannot transfer to another species. As a result, if one plant is destroyed by disease, the entire plot is likely to survive.

In contrast, monocrop agriculture and ranches have been disastrous both for the development interests and the local peoples who are displaced. Both monocropping and ranching quickly deplete the soil and leave arid savannahs where tropical rainforests once stood. Tropical diseases can easily sweep through monocrop farms and destroy an entire harvest. Farmers often lack the financial resources to buy pesticides, and pesticides that are effective in temperate climates may be no match for the myriad pests of tropical climates. Incidentally, the argument against monoculture applies to temperate cultures as well. In upstate New York and many other areas of the United States, small farmers find it hard to stay in business. However, I recently met a farmer from upstate New York who explained how he was able to turn a profit on his farm within a year through a highly diversified program of crop and livestock production.

Studies of indigenous agriculture and forest management provide the evidence that is needed to challenge the large-scale development schemes of third world governments and first world aid agencies. Bankrupt third world governments that face population pressures often see their tropical forests as a source of quick revenue and a safety valve for population growth. However, strategies that lead to deforestation ultimately only impoverish third world countries, for they destroy one of their most valuable resources. Enlightened governments and aid agencies now look to indigenous peoples and other forest peoples not as savages needing to be civilized or relocated but as the rightful owners of the forest and as teachers whose local knowledge can provide the basis for successful natural resource management. From this perspective the study of ethnoknowledges can be tapped to support the struggles for cultural survival and social justice of indigenous peoples throughout the world.

Non-Western Medicine

As Western medicine spread throughout the world during and after the colonial period, it generally coexisted with local, non-Western systems of medicine. Non-Western medicines, sometimes called ethnomedicines and ethnopsychiatries, ranged from systems of countersorcery and exorcism to the highly developed, literate medical traditions of China and India. Advocates of Western medicine have usually adopted a straightforward, acultural approach to their non-Western rivals. They argue that Western medicine gradually will replace all the other medicines, because Western medicine is simply more effective. In contrast, a cultural approach can lead to a more sophisticated understanding of the various ways in which Western and non-Western medicines are related. I distinguish several ways of thinking about Western and non-Western medicine from a comparative, cultural perspective: institutional relations, therapeutics, concepts of etiology, concepts of illness and disease, and the incidence and translatability of illness and disease across cultures.

Institutional Relations and Medical Pluralism

At first glance, some convincing arguments support the commonsense evolutionary perspective that sees non-Western medicine as simply an inferior and outmoded medicine. For example, it is difficult for anyone to quarrel with Western medicine's success at controlling many infectious diseases. Although antibiotics and immunization have their share of shortcomings, they have also saved many lives. Likewise, Western medicine is also very effective at saving lives in cases of traumatic injury and even in many chronic, degenerative diseases.

However, a large number of studies in the medical anthropology literature has demonstrated that non-Western medical practices are also effective, particularly outside conventional Western strongholds such as treatment of infectious disease and major injury. For example, non-Western herbal remedies are often pharmaceutically active, and non-Western healers often are adept at setting bones, reducing pain and anxiety, and providing suggestions that help the healing process. Furthermore, because non-Western healers have a better understanding of their communities, they may end up being much more effective at healing psychological and social problems than medical practitioners who are trained in Western medicine and are unfamiliar with the local culture.

Because non-Western medical traditions are often embedded in ritual practices that are closely woven into an entire social order, people are likely to continue practicing their local medicine even when Western medicine becomes available to them. As a result, in many non-Western settings the prognosis of Western-trained clinicians that non-Western medicines will gradually wither away has turned out to be wrong. Instead, a situation of medical pluralism tends to emerge. Today most medical systems in the world are complex mixtures of Western and local or non-Western medical systems. The situation is true not only in poor, non-Western countries but also increasingly in rich, Westernized countries.

The question of colliding or plural medical systems—medical pluralism—is of general interest for the cultural studies of science and technology because it provides a framework for the relationship between cosmopolitan science and technology on the one hand and local ways of knowing and doing on the other hand. In other words, medical pluralism can provide a model for studying what might be called techno-scientific pluralism.

Although the phenomenon of medical pluralism is now well-documented, some social scientists have called into question the assumptions implicit in the word *pluralism*. They argue that while it is true that most medical systems throughout the world are complex or multiple, the term *pluralism* suggests that the relations between Western and other medicines are benign and untroubled by asymmetries of power. The critics also argue that Western or cosmopolitan medicine often has the backing of the state, which historically was usually controlled by colonial powers. For this reason the terms *medical hegemony* or *medical domination* are sometimes preferred to signal how Western medicine usually is embedded in a system of colonial or neocolonial power relationships.

Consider, for example, the case of the introduction of vaccination into India, as described by the anthropologist Frédérique Apffel Marglin (1990). Traditional India already had a fairly sophisticated system of smallpox control through the variolators associated with the cult of the smallpox goddess Sitala. Variolation involved taking pus from infected people and pricking the skin of uninfected people. The procedure is almost the same as vaccination, except that vaccination uses cowpox. The advantages of vaccination are that the inoculated person is not contagious and the mortality rate is lower. However, the British introduced vaccination by outlawing variolation, and they insulted the goddess Sitala by talking of smallpox as an absolute evil that should be eradicated. As a result, they encountered unnecessary resistance that could have been avoided if they had worked in partnership with variolators. Their decision not to work in partnership reflected their sense of superiority over the Indians, and in turn their sense of superiority rested in large part on their belief that their knowledge, technology, and medicine were better.

Apffel Marglin's study shows how Western medicine was introduced as part of a colonial program to control local communities by displacing indigenous medicine and, with it, alternative indigenous authority in medicine and religion. That program of displacement in medicine was in turn part of a larger program in which indigenous knowledge and technology were often closed down (as in the case of textile production), even as Western powers absorbed and borrowed local knowledges. A more benevolent national or colonial government might sanction systems that allow Western and local medicine to coexist. For example, the anthropologist A. P. Elkin (1977) discusses a case in which an Aborigine was trained in Western medicine and was able to practice with recognition as both a "Blackfellow Doctor" (shaman) and "Whitefellow Doctor" (medical doctor). Elkin argues that the best system would be to give Native shamans—that is, medicine men and women—the opportunity to incorporate cosmopolitan medicine into their practices. As experts in both cosmopolitan and local medicine, the shamans would then be in a good position to negotiate local and cosmopolitan knowledges and therefore to understand and treat their patients better. This locally based approach would also be likely to create some interesting new blends of local and cosmopolitan medical systems.

The cases of Indian variolators and Australian shamans show how the relations between Western and traditional medicine are part of broader power relations. The cases also show how the relationship can vary depending on the level of cultural sensitivity of the Western medical community. Another factor that affects the relationship between medicines is the rural/urban dimension. In larger cities where there are people from different regions or even different countries and ethnic groups, the system of medical pluralism tends to resemble a competitive marketplace with many products rather than a simple dyadic relationship of Western to local medicine. Although Western medicine usually maintains an upper hand through control over hospitals, access to government funds, rights to insurance, and favorable legal regulations, in many cases other medical traditions have acquired a substantial amount of financial and institutional power as well as legal recognition.

Only in recent years have the state and medical profession in the United States come to terms with the reality of the vibrant medical pluralism of the large urban areas. In the early 1990s the American government began to sanction medical studies of the efficacy of alternative therapies through the National Institutes of Health. Likewise, in 1993 the *New England Journal of Medicine* published results of a national survey that revealed that over 34 percent of the American people had resorted to at least one unconventional therapy in the past year. Still to be recognized is the extent to which alternative medicine is also non-Western medicine. In what might be viewed as part of the great boomerang or countercolonialism of the global society, in several developed Western countries there are millions of middle-class people of European descent who are experimenting with therapies of non-Western origin. As the societies of the old colonial centers—Europe and nations with predominantly European populations—are becoming more multicultural and globalized, so are their medical systems.

Non-Western Therapies in the United States

In addition to institutional issues, the question of medical pluralism involves complex therapeutic relationships. Non-Western therapies are a diverse group, and practices that might be classified from a Western perspective as "therapies" often overlap with categories such as religion, diet, exercise, or relaxation. That should come as no surprise, because the non-Western medicines come from cultures that cut up the world into different categories than those of the West after the Scientific Revolution. As a result, it might be better to use the label *non-Western health practices* rather than *non-Western therapies*. Table 7.1 provides a brief list of non-Western health care practices that are commonplace in the United States. From the list it is evident that non-Western medicine in the United States generally refers to East or South Asian health practices. The one popular exception is shamanism, which draws on Native American religion and to some extent indigenous traditions from other parts of the world. Other non-Western therapies exist, but they tend to be localized in specific ethnic populations. For example, in Latin-American neighborhoods of northern and eastern American cities, Santería and related African religious practices from the Caribbean are fairly common. The Santeros often work with herbal remedies, or they perform exorcisms, spells, prayers, and countersorcery rituals.

Although a variety of non-Western therapies is available, there is a division of therapeutic labor. In the United States people with an infectious disease or a severe traumatic injury are likely to use the standard

TABLE 7.1

Some Non-Western Therapies Currently in Use in the United States

Acupressure	Similar to acupuncture but uses finger pressure instead of needles
Acupuncture	A procedure that involves inserting needles at special points on the body surface that correspond to a system of subtle-energy meridians that run up and down the body
Ayurvedic	An ancient Indian medicine based on a humoral theory similar to that of ancient Western medicine
Macrobiotic	A vegetarian diet system of Japanese origin that uses grains, fruits, and vegetables
Meditation	A way of calming the mind and relaxing by focusing attention on a single idea, visual point, or set of words
Reiki	A therapy of Japanese origins based on the transfer of a universal healing energy in a hands-on procedure
Shamanism	Healing and psychologies of personal development achieved by inducing a "shamanic voyage" through dancing, drumming, or chanting
Shiatsu	A Japanese massage practice that is a variant of acupressure
Tai-chi	A Chinese system of exercise that involves moving the body slowing through a series of postures
Yoga	The Indian science of the mind that includes meditation or hatha yoga, which focuses on body postures

therapies of physiological biomedicine such as antibiotics or surgery. Likewise, members of most ethnic groups who are suffering from a severe mental disorder are likely to end up in the hands of a psychologically trained professional who relies on psychoactive medication and/or psychotherapy. Although the term *biomedicine* is sometimes used as a general category for all Western-style therapeutics, I shall use it here in a more restricted sense to refer to physiologically based or body-oriented therapies in contrast to psychotherapies. Together, biomedicine and psychotherapies form the two main branches of official Western medicine. Non-Western therapeutic practices proliferate for the kinds of physical and mental dysfunctions that official biomedicine and psychotherapies do not recognize or cannot heal, such as some degenerative diseases or chronic pain.

One aspect of the role of non-Western therapies is that they tend to take much more seriously the idea of health and the active role that one must play in maintaining it. Although official medicine increasingly recognizes the importance of diet and exercise in maintaining health, non-Western practices tend to take stronger positions. For example, it is not enough to lower one's cholesterol level; the controversial advocates of macrobiotic diets would advocate ridding the diet of all animal products and processed foods. Likewise, it is not enough merely to get enough rest, exercise regularly, and avoid a high-stress lifestyle; the various practitioners of meditation, yoga, massage, and Tai-chi would suggest that one should also learn to still the mind and relax the body.

In addition to pushing the boundaries of the idea of health and preventive medicine, some non-Western health practices also serve as adjuncts to biomedical therapies. In the case of degenerative diseases such as cancer, the efficacy of biomedicine is very limited, and the side effects of its primary therapies—which until recently were limited to chemotherapy, radiation, and surgery—are often debilitating. In that situation patients may turn to non-Western practices as adjunct therapies they hope will strengthen their immune system or help trigger immunological defenses. Likewise, patients sometimes turn to non-Western practices to help them overcome pain through procedures of mental disciplining or bodily relaxation. Finally, in cases where patients have lost hope with biomedicine, they may turn to non-Western practices for the cures they could not find in biomedicine. Obviously treatment of last resort occurs most frequently in cases of chronic pain and degenerative diseases for which biomedicine has no available solution.

In the area of psychology and psychiatry there is a similar division of labor. For example, just as non-Western medical practices can serve as a more extreme form of preventive medicine, so many of these practices might be viewed as a more extreme form of the psychologies of personal growth and development (a kind of preventive psychotherapy). In this context, people see non-Western practices less as therapies meant to solve particular dysfunctions and more as methods for achieving a greater sense of personal fulfillment. Non-Western practices often mix freely with conventional Western psychologies of personal growth, and the non-Western practices often find a welcome home in the seminars and weekend retreats of humanistic psychologists, Jungians, and transpersonal psychologists. Those groups serve as bridges between the Western psychological tradition and non-Western spiritual traditions.

Non-Western therapies can also substitute for or serve as adjuncts to Western psychotherapies in the case of people who wish to remedy a particular psychological dysfunction. Classical Western psychotherapy emphasizes a talking cure that is achieved through intellectual insight, emotional expression, learning, conditioning, and/or the recall of repressed memories. In contrast, non-Western therapies share a cultural space with the less mainstream Western psychotherapies that emphasize the role of body energy as a key element in the psychotherapeutic process. Thus, many therapists who draw on non-Western practices integrate those practices into heterodox Western psychotherapies of bioenergy (a term I use to cover a range of beliefs in subtle energy fields). Those therapies are heterodox because they rely on concepts such as vibrations, auras, and human energy fields that most of the scientific community rejects as lacking convincing empirical evidence. Body-oriented therapies are also different from the mainstream of psychotherapy because they maintain that the solution to a psychological dysfunction cannot be achieved wholly at the mental level. Instead, the traumas, tensions, and anxieties of the dysfunction have been inscribed on the body, and it is the task of the therapist to help release those points of tension.

Non-Western therapies therefore occupy a number of positions in a pluralistic medical system. I have distinguished four main areas in the medical division of labor: preventive biomedicine, therapeutic biomedicine, personal growth psychologies, and heterodox psychotherapies. Non-Western therapies occupy a more extreme position in the domain of preventive biomedicine and personal growth psychologies, and a position of adjunct (and sometimes replacement) therapies in the area of therapeutic biomedicine and psychotherapies. Regarding psychotherapies, non-Western ideas and therapies have already merged to some extent with their Western counterparts. For example, therapists influenced by Carl Jung are likely to be open to a number of non-Western ideas and practices, just as are those who believe in bioenergy and the need to anchor psychotherapy in body energies.

Other Etiologies and the Encounter with Biomedicine

So far, non-Western therapies have had little influence on biomedicine per se. Biomedical researchers may accept the idea that non-Western practices can help patients to relax or lower pain, and they may even accept some cases of "cures" by explaining them as examples of the effects of suggestion, placebo, spontaneous remission, or psychosomatic healing, where such effects are believed to be possible. Likewise, pharmaceutical companies have invested substantial amounts of research funding in the exploration of possible therapeutic effects of non-Western herbal medicines. However, when it comes to accepting the theories that accompany non-Western medical systems, Western or cosmopolitan researchers in medicine, psychiatry, and psychology tend to draw the line. Nevertheless, in a few cases the movement of counter-colonial medicine is beginning to have research implications in the biomedical community.

One example is the attempt to test acupuncture points. The standard scientific explanation of the analgesic (pain-reducing) effects of acupuncture is that the insertion of the needles works as a placebo or suggestion as in some forms of hypnosis, or that in a more direct way the needles stimulate the body to produce endorphins that lower pain. Although most researchers recognize that acupuncture can in some cases reduce pain through hypnosislike or placebolike effects, the endorphin hypothesis has been weakened by studies that have shown there is no relationship between plasma endorphin levels and pain. A more profound attempt to situate acupuncture within the framework of an altered Western medicine is the work of the medical researcher Robert O. Becker and his colleagues, who have found that the electrical resistance of about half the acupuncture points was different from control points. Again, however, the results are not conclusive because, as far I was able to ascertain, the studies have not been replicated in other laboratories and double-blind protocols were not used.

Another area in which non-Western medical systems have entered into biomedical research involves the use of meditation and related psychological practices in the treatment of cancer. The Australian psychiatrist Ainslie Meares published a number of papers in the late 1970s and early 1980s in which he argued that a treatment protocol of twenty sessions of intensive meditation led to reduced anxiety, depression, discomfort, and pain in about half of his cancer patients. In some cases in which patients did not use conventional treatments but did use meditation, he found regression of cancer. However, the results remain in the category of anecdotal research. For most medical researchers that type of research may be useful because it suggests avenues that may be worth pursuing in controlled studies. The most convincing controlled experiments involve quantitative human studies in which patients are paired off before the protocol is applied (known as prospective, clinically controlled trials).

The two examples show that although non-Western therapies may be flourishing in Western societies, they have not had much impact on biomedical research. When non-Western therapies do make their way into biomedical research projects, they are usually shorn of their non-Western theoretical frameworks and reinterpreted in terms of current biomedical and psychological knowledge. Thus, there is a limit to the extent to which medicine is becoming multicultural. Although the medical system as a whole may be becoming increasingly pluralistic and multicultural, like it or not the theorizing of the universities and large research organizations remains largely impervious.

The Cultural Construction of Illness and Disease

In addition to institutional relationships, therapies, and theories of etiology, the interaction between Western and non-Western medicine also involves understandings of illness and disease. The distinction between *illness* and *disease* has a specific meaning in the medical anthropology literature that stems in large part from the work of Harvard University doctor and anthropologist Arthur Kleinman. In *Patients and Healers in the Context of Culture*, Kleinman defines the terms as follows: "*Disease* refers to a malfunctioning of biological and/or psychological processes, while the term *illness* refers to the psychosocial experience and meaning of perceived disease" (1980:72). The framework set up an opposition between the authority of doctors trained in Western medicine and psychiatry and that of patients. The latter draw on local, non-Western medicine and/or their understanding of official medical knowledges, which they synthesize as their own concepts of illness (50ff.).

This formulation of the illness/disease dichotomy has encountered several criticisms. From the quarters of critical medical anthropology, some have argued for an alternative framework for which the key rubric is *sickness,* a term used to signal the way illness and disease involve power relations between doctors and patients. The main problem is that the illness/disease dichotomy can be used to mean that illness is just the patient's understanding of what is going on, whereas disease is what is really happening. That interpretation, however, obscures the way in which the diagnosis and etiology of professionals is every bit as much a construction as that of the patient. To some extent Kleinman recognizes the point in his discussion of illness and disease, although the point seems to get lost when others employ the dichotomy. Kleinman argues that both disease and illness exist "as constructs in particular configurations of social reality" and that they express "different interpretations of a plural clinical reality" (1980:73). Thus, whereas the illness/disease dichotomy can be useful, the STS perspective that disease is as much a social construction as is illness should be made a part of the analysis.

Notwithstanding some of the confusions, medical anthropology has greatly contributed to understanding the patient-practitioner interaction. Medical anthropologists have shown how the process of illness/disease construction is embedded in a pluralistic medical system in which patients and doctors must negotiate each other's constructions of the illness/disease. By viewing illness/disease construction as a negotiated process, medical anthropologists have pointed to a number of "silent boundaries," to use the phrase of the anthropologist Karen Pliskin in her study of illness/disease construction among an ethnic minority group in Israel. Pliskin describes those boundaries as

> the various social and cultural elements and circumstances which, unbeknown to patient and practitioner, impede successful therapy: communication differences; cultural concepts of emotions, expression, sickness; ethnicity and stereotyping; medical praxis. (1987:8)

A culturally naive clinician might not even be aware of the many miscommunications and misunderstandings that occur in the clinical encounter; a culturally sophisticated clinician now has resources for better understanding the clinical encounter, especially when it involves ethnic and gender differences between patient and doctor.

Viewing medical/psychiatric definitions of disease and mental disorder as socially constructed does not mean that they are purely imaginative devices that have no correspondence to what is out there in the world. Rather, it means only that the current understanding of what constitutes a disease or a psychiatric disorder at any given time is a product of the history of a scientific and medical community. Furthermore, understandings are subject to change over time. As knowledge of psychosociobiological processes and etiologies change, so do classifications and etiologies.

The social construction of disease becomes most evident—and politicized—in cases such as AIDS. In such cases definitions of when the disease begins affect civil rights as defined in terms of access to treatments and insurance. The socially constructed nature of medical knowledge is also very evident in debates over psychiatric disorders, which have frequently become highly politicized. In the United States debates often focus on whether or not a given category of psychiatric disorder should be included in the psychiatric classification system known as the *Diagnostic and Statistical Manual* (DSM). The system itself has undergone many revisions, and by the early 1990s the fourth version was in preparation.

The terminology and understandings of psychiatry have changed dramatically since the nineteenth century and even the first half of the twentieth century. The most notorious example of older categories is hysteria, which is derived from the Greek word for *uterus*. The ancient Egyptians believed that the uterus could move around the body and that the "wandering womb" was responsible for the affliction. The Egyptians fumigated the woman's genitals to attract the uterus back to its "right" place, or they gave the woman unsavory potions to drive it back down. The Greeks developed the Egyptian theory and treatment by arguing that the illness occurred when women remained "barren" too long after puberty. The solution was to get married and have children, and as a result the illness came to have a moral meaning as a sanction against women who did not accept marriage and motherhood. Closer to our time, doctors sometimes treated "hysterical" women by performing a hysterectomy.

It was only with Freud and the first generation of clinical psychologists that the psychological bases of hysteria became more evident. During that time the diagnosis of hysteria was broadened and it became more common to speak of hysterical symptoms in men as well as women. Although Freud's studies of hysteria were limited by his own patriarchical attitudes, his methods made possible subsequent work that reinterpreted hysteria increasingly as a form of resistance for abused women whose means of protest were limited by conservative female sex roles. Today hysteria has become one of the categories of diagnosis that the American psychiatric profession has dropped from its classification scheme because of fuzziness and sexist bias.

The his-story of hysteria provides an example of a mental disorder that has basically disappeared in recent years because of changes in both symptom frequency and diagnostic preferences. In general, there has been a progressive tendency to eliminate some of the most sexist and heterosexist categories of psychopathology from official diagnostic manuals. For example, from 1952 to 1973 the American Psychiatric Association listed homosexuality as an illness. Although Freud argued against classifying homosexuality as psychopathology, in the more puritanical American environment some leading psychoanalysts ignored his position. In 1973 activists (including some psychoanalysts) finally succeeded in removing homosexuality from the DSM. Subsequent attention focused on sexist bias in diagnosis patterns for and assumptions behind several of the commonly used labels for personality disorders, such as "histrionic," "dependent," and "self-defeating personality disorder." Regarding the "dependent" personality disorder, some have suggested instead a "restricted" personality disorder for men under the spell of machismo.

In some cases disorders do not disappear as much as change names, symptoms, and etiologies. Depression, for example, was originally known among the ancient Greeks as *melancholia,* a chronic disease with symptoms of sadness, fear, misanthropic attitudes, delusions, gastrointestinal problems, and a general sense of being tired of life. During the medieval European Christian period a related condition known as *acedia* received great attention. Often considered a sin, acedia was sometimes associated with sloth, carelessness, negligence, and other negative moral characteristics. The modern illness of depression emerged as the moral overlays of acedia were dropped. The term current at the beginning of the twentieth century, *neurasthenia,* was associated with a disorder that resulted from weakness or exhaustion of the nervous system. Its symptoms included headaches, insomnia, dizziness, fatigue, weakness, anxiety, irritability, and other localized nonorganic pains. Notice that neurasthenia lacked the moral quality of acedia, and it also lacked several of the symptoms as well as the chronic status of melancholia. Later, as psychological understanding increased both among the medical profession and the general public, the physical symptoms tended to be replaced by more psychological self-reports, and modern *depression* emerged. Still other reclassifications are likely as new drugs that work on specific neurotransmitters, such as Prozac, lead to regroupings of symptoms and disorder categories.

Some psychiatric disorders also disappear only to reappear later. For example, anorexia was recognized in earlier centuries as anorexia mirabilis, a condition that can be traced to the fourteenth-century Saint Catherine of Siena and the general Christian tradition of saintly women who were reputed to live only on the Eucharist. As the historian Joan

Brumberg discusses in *Fasting Girls*, the tradition continued into the nineteenth century in the form of "miraculous maids," who were also considered saintly and were reputed to live without nourishment. However, in the twentieth century a somewhat different form of anorexia, anorexia nervosa, became common among young women who grew up in families in which parents use food as a form of discipline and in cultures where advertisements teach them that thin is beautiful. By the middle of the twentieth century, anorexia nervosa had reached nearly epidemic proportions in the United States, and so much attention was given to the disorder that in 1985 a separate diagnosis was created for bulimia, a syndrome of binging and purging.

Another example of a disorder that went through highs and lows is multiple personality. The disorder flourished during the late nineteenth and early twentieth century, but it fell into disrepute after the early twentieth century when hypnosis fell out of favor. For decades patients with similar symptoms were diagnosed as schizophrenics or hysterics. Blackouts or dissociative states were frequently attributed to temporal lobe seizures rather than to hypnosislike amnesias or dissociations. Furthermore, the diagnosis of schizophrenia made it possible for patients to receive insurance, whereas a diagnosis of multiple personality was less likely to be accepted. However, during the last decades of the twentieth century hypnosis research and therapy reemerged, and psychiatrists also became more aware of child abuse as a cause of mental illness. Multiple personality again became popular as a diagnosis, and the category made it back into the official diagnostic manual in 1980. In 1984 the International Society for the Study of Multiple Personality and Dissociation was founded. The new understanding of multiple personality attributes the disorder to child abuse that results in the "dissociation" or splitting of painful memory clusters into separate personalities.

In short, to say that Western notions of mental disorder are socially constructed implies more than saying that social factors have shaped those ideas. The concepts, etiologies, symptoms, and terms change radically over time, even to the extent that some disappear from the lexicon. Furthermore, the very definitions and debates over inclusion and exclusion are deeply reflective of the elite, heterosexual, European, male professional culture from which they emerged. As the medical profession opens up its doors to other groups—or no longer has the power to keep the doors barred shut—previously uncontested knowledge becomes highly politicized.

The Cultural Distribution and Translation of Illness

As for the medical construction of disease and disorder, a cultural perspective can also provide a framework for understanding aspects of
their epidemiology. For example, it is well-known that the distribution of
diseases and psychiatric disorders is far from uniform over the world,
and that the epidemiology is shaped by social practices as well as environmental factors. Schizophrenia and anorexia, as well as cancer and
heart disease, are much more prevalent among the wealthy, developed,
Western countries. Likewise, some diseases may be more or less limited
to certain areas of the world, such as infectious tropical diseases, but
even within those limited areas there are radically different distributions
due in part to the varying cultural practices or differential access to sanitary systems and health care.

In addition to playing a role in the social shaping of the incidence
of the disease, local medical practices shape the symptoms and understanding of the disease itself. For example, as Kleinman discusses, in
the People's Republic of China depression is frequently diagnosed as
neurasthenia. The older term fits well with the emphasis on an organic
approach that in turn fits better with the materialistic communist ideology. Furthermore, patients in China tend to somatize more, that is, they
tend to experience their symptoms through their body. In contrast, in
the West, where psychology is much more a part of the popular culture,
patients tend to experience and express their symptoms at a psychological level.

The case of the People's Republic of China is interesting because it
shows how the university-trained medical profession that is part of a
transnational medical community nevertheless operates not only with a
different understanding of depression but also with a different set of symptoms. A problem of translation emerges. Is Chinese neurasthenia really
Western depression with a slightly different set of symptoms, or should
we think of them as two related but different psychiatric disorders?

The problem of translation is even more evident in the case of traditional classifications of non-Western illnesses. For example, the psychiatrist Raymond Prince describes the traditional classification system of
the Yoruba of West Africa as having a number of categories that do not
always map in a single and one-to-one way onto Western psychiatric disorders. *Asinwin* refers to an acute (short-term) psychotic episode in
which the patient may commit suicide or murder. Prince, writing in the

early 1960s, mapped *asinwin* onto a number of Western psychiatric categories at that time: acute schizophrenic episode, mania, catatonic excitement, and agitated delirium. Likewise, he wrote that *aiyiperi* is "a complex concept that includes hysterical convulsive disorders, posturings, and tics, as well as psychomotor seizures and probably tetanus" (1964:88). Another illness, *ori ode* (hunter's head), involves a severe headache that feels like ants crawling inside one's head or like an anvil is pounding inside the head. The obvious candidate for translation is the migraine headache, but the Yoruba attribute the illness to sorcery, and it therefore takes on specific symptoms related to sorcery beliefs. Thus, it might also be seen as a conversion disorder based on belief that one has been ensorcelled.

It is possible, then, to translate non-Western categories of illness into Western categories of disease and psychiatric disorder. However, some localized folk illnesses seem to have a very specific set of symptoms (that is, they form a "syndrome") that lacks any comparable counterpart in Western/cosmopolitan psychiatry. The term *culture-bound syndromes* emerged in psychiatry and anthropology to describe folk illnesses that were originally assumed to be located only within a specific culture or group of cultures.

An example of a culture-bound syndrome is *koro*. In East and Southeast Asia men are sometimes struck with the sudden fear that the penis is going to retract up into the body and thus kill them. (There are some reports of a similar condition for women's breasts, but it is rarely reported.) The condition is widely recognized and accepted in the cultures of the region, and there are standard explanations of why it occurs and what to do to stop it from happening. There are also a number of well-documented cases indicating that *koro* attacks do occur among males in those populations; in other words, it is a syndrome rather than merely a belief. The question of whether or not the penis can actually disappear into the body is moot. As some researchers have argued, it is possible that the anxiety of penile retraction leads to decreased blood flow to the penis, which in turn causes a perception of penile shrinkage, which creates more anxiety, which leads to decreased blood flow, and so on.

Koro has been classified as a folk illness that is simply a panic attack combined with perhaps an obsessional neurosis. However, over the years, every now and then a study in the psychiatric literature pops up that contributes growing evidence that korolike symptoms occur among adult males in very different cultures from those of Southeast and East

Asia. There are now reports of koro attacks in an American schizo-phrenic, a French Canadian, an Anglo-Saxon Canadian, a Yemenite, a Georgian Jew, a Nigerian, a Londoner and various other Englishmen, not to mention epidemics in Singapore and Thailand. The possibility emerges that koro may not be culture-bound at all, but instead a mental illness that has, to use the phrase of the psychiatrist Ronald Simons, a "neurophysiological shaping factor." Of course, the alternative possibil-ity is that given the perhaps overrated importance of the organ, the same idea happens to occur to men throughout the world, but it is only in cer-tain cultures that the idea becomes elaborated into a folk illness.

A number of other folk illnesses have received attention as discrete, culture-bound syndromes. For exampie, *latah* is a condition common in Southeast Asia (not necessarily viewed as pathological) in which a per-son is a hyperstartler who tends to lose control after being startled. After being startled *latahs* often yell out obscenities and uncontrollably mimic people around them, often to the amusement of people watching. Other culture-bound syndromes include *amok*, another Southeast Asian syndrome, in which people commit a mass suicide-murder; *piblotoq*, found among Eskimos who sometimes are seized by the dangerous urge to take off their clothes and run naked in the snow; and various sleep paralysis conditions in which sleeping people are visited by spirits and find themselves unable to move. The debate on culture-bound syn-dromes has focused on the extent to which similar syndromes in differ-ent cultures are really in some sense the "same," and whether or not there may be biological factors that could account for apparent cross-cultural similarities.

The debate on culture-bound syndromes has two major implications for Western or cosmopolitan psychiatry. One is that it challenges health professionals to rethink critically their own categories of mental disor-der. As discussed above, there is already an ongoing debate on the cate-gories, etiologies, and treatments of mental disorder within the various medical professions of the world. In the United States the debates tend to emerge from political or moral considerations, such as questions regarding the politics of hysteria or homosexuality as psychiatric disor-ders. The comparative perspective offered by the study of culture-bound syndromes makes it possible also to ask to what extent some of the Western psychiatric disorders are products of a specifically Western cul-tural setting rather than of presumed biological processes.

For example, some proponents of biological theories of anorexia nervosa believe that it results from a defect of the hypothalmus, which controls metabolism, temperature, respiration, and food and water intake. However drug treatments have not been more successful than a placebo, and anorexia also follows a "development gradient": the disorder is found in Western or Westernizing countries, including non-Western ethnic groups within those countries, and it is rare in Africa, Asia, and the Middle East. Anorexia also increases in prevalence over time, with cases on the rise in recent decades. It is found among the middle and upper class, overwhelmingly among women, and especially among intelligent and ambitious women. Athletes, dancers, dietitians, models, and racing jockeys are most prone to be anorexics. Given this epidemiological profile, it should be no surprise that debates are now emerging about the nature of anorexia. Perhaps it is not a biological disease with an etiology shaped by cultural factors, and instead it is a culture-bound syndrome that appears in cultures where food and parental love are deeply interwoven, and thinness is exalted in a cult of youth and beauty that is the product of continuous media bombardment.

What happens when culture, and the multicultures of the global society, are let into the conceptual frameworks that guide the classification systems of international psychiatry? By the 1990s the medical profession was in the ninth version of the system of the World Health Organization (ICD-9) and in transit from the third to the fourth version of the American psychiatrists' standardized system of classifications (DSM-IIIR). What will ICD-23 or DSM-17 look like? One possibility is that as more clean drugs (specific to one neurotransmitter) such as Prozac appear, psychiatry will become increasingly biologized, and psychiatric disorders will come to be understood more and more as biological diseases. In this scenario culture will be eliminated (at least at the explicit level) and many of the current diagnostic categories will pass into historical oblivion or popular usage as folk illnesses.

However, it is also possible that psychiatry will increasingly think of all aspects of mental illness—epidemiology, etiology, and treatment—as a biopsychocultural process. It would be possible to rethink the entire classification system by starting from all the folk illnesses of the world (the emic level), including commonly accepted Western categories of mental illness. They could be reclassified according to symptoms into a genuinely international system (the etic level). Such a system would probably be of some use to health professionals, who work increasingly with patients from diverse cultural backgrounds. It would allow them to understand better what patients mean when they describe sets of symptoms, and it would also allow them to design culturally sensitive treatments.

Conclusion

The study of the various ethnoknowledges—from ethnobotany to ethnopsychiatry—may appear to be a quaint corner of the academy in which non-Western ways of knowing and doing are catalogued and analyzed. However, as I have argued, studying the ethnoknowledges can also be a reformist project at a number of levels. By showing that non-Western knowledges, technologies, and medicines are often coherent and elaborate, an intellectual resource emerges that can be used to resist the ideology of development interests who wish to impose unwanted Western knowledges and technologies in the name of civilization. Furthermore, by showing that non-Western ways of knowing and doing are often efficacious and in some cases superior to Western or cosmopolitan alternatives, it is possible to build a resource base for critiquing and contributing to changing development projects.

At another level, the study of ethnoknowledges makes it possible to put into question the universalistic assumptions of cosmopolitan science and technology. Some of the basic categories of animal, plant, and disease/disorder classification may be confirmed at a general level through the study of convergences, but alternatives also emerge. For example, Western medicine and psychiatry can be put in dialogue with other notions of the body that reveal the diseased body to be a social as well as biological phenomenon. It may even happen that non-Western notions of bodily energies will eventually become translated into Western theories of electrobiology and related new fields.

8 | Cosmopolitan Technologies, Native Peoples, and Resistance Struggles

The unfortunate truth of the current situation is that the relationship between Western and indigenous knowledge and technology is far from benevolent. Economic inequality, resource depletion, and overpopulation in the national societies are among the moving forces behind the continued colonization of indigenous lands. Although there are some model cases such as ethnomathematics education and mixed medical projects, more often Western science and technology are tools in the ongoing process of the destruction of indigenous land and culture. That process has only accelerated with each passing decade. From the perspective of Native peoples the immediate problem is devising strategies of how best to resist unwanted incursions on their territory and way of life. In this chapter I explore the complicated nexus of technology, culture, and resistance in the context of indigenous peoples.

Indigenous peoples are often, to use the phrase of the anthropologist John Bodley (1982), the "victims of progress." The phrase is a tricky one because the word *victims* implies a passivity that, as I shall show, is by no means an accurate description of the resourcefulness of Native peoples in their resistance struggles. Furthermore, Bodley uses the word *progress* in an ironic way, for it is a progress that Native peoples frequently neither control nor want, and it is also a progress that has become associated with ecologically unsustainable social orders in the developed countries. Growing sectors of the national societies are coming to question the traditional value of progress, especially when it involves building roads, ranches, mines, hydroelectric dams, and so on in areas of the world that Native peoples have claimed as their own for generations or even centuries. The issue is more than a moral one of human rights and territorial integrity; increasingly, members of the great technoindustrial civilizations are beginning to see that Native peoples have a great deal to teach us about ecologically sustainable ways of life.

In any encounter between two societies technology usually plays a key role. The point is obvious in today's high-tech age. For example, in the Persian Gulf War of 1991 the world watched as the superior technology of the Western allies devastated an often helpless Iraqi military and civilian population. However, the point was also true in earlier times. The tiny army of England's Henry V was vastly outnumbered by the French, but the English managed to win largely due to the technological advantage of their crossbows over the French long bows. (The crossbows, incidentally, are another example of a Chinese innovation that made its way to Europe.) In the invasion of the New World the Europeans had even greater military superiority over the indigenous peoples, and small numbers of Europeans were able to gain control of entire civilizations such as the Incas and the Aztecs. Although Spaniards such as Pizarro and Cortez relied on trickery and local insurrection as much as on their horses and canons, they nevertheless used their superior technology to give them a strategic advantage. In today's world of computerized air warfare the technological gap between national and Native societies is even greater than at the time of the invasion of the New World. Native peoples today occasionally resort to arms, but their resistance struggles are likely to be reduced to guerrilla warfare and terrorism rather than direct confrontation with the generally much better equipped national militaries.

Although conquest and invasion have been characteristic of the relations between the national societies and Native peoples throughout the world, not all relations between Native and national societies can be classified in such negative terms. In many cases the two types of societies coexist peacefully through trade relations and other forms of relatively benign contact. Still, even when contact does not involve the wholesale expropriation of indigenous peoples from their lands, cosmopolitan technologies diffuse into indigenous societies. The effects of those technologies, even when Native peoples (or segments of Native societies) want them, are hard to predict, and sometimes the new technologies lead to major social disruptions. In this chapter I will consider some of the frameworks for examining Western technologies and indigenous peoples, and various strategies for helping indigenous peoples obtain more control over their land, lives, and relationship with national societies and technologies.

The Social Impact of Technology Approach

Theoretical Frameworks

In cultural anthropology an influential set of frameworks for analyzing the role of technology in the relations between Native peoples and cosmopolitan societies is derived from debates over theories of cultural evolution. In the 1940s the anthropologist Leslie White proposed that cultural evolution was influenced by the amount of energy harnessed per capita per year and "the quality or efficiency of the tools employed in the expenditure of energy" (1949:368). Although not everyone accepted White's theory, it had the advantage of focusing attention on questions of how and why cultures change and on technology and energy consumption as key factors in culture change. In the next decade the anthropologist Julian Steward developed a related theory that linked culture change and evolution with adaptation to the local environment, and evolutionary theory in anthropology has continued to remain alive today through studies of cultural ecology.

Discussions of the nature of culture change and evolution helped lay the theoretical groundwork for the study of the impact of Western technologies on Native communities. In the politically conservative climate of the United States during the cold war, the discussions also helped make it possible to air ideas that were similar to those proposed by Karl Marx a century earlier. With these discussions on technology and culture change in the background, anthropologists had a framework for analyzing the impact of new technologies on the entire fabric of the culture: its social organization, political structure, religion, and so on. As a result, the first generation of studies of technology, indigenous peoples, and culture change examined the question in terms of the impact of the diffusion of cosmopolitan technologies on local cultures.

One guiding heuristic that emerged in the social impact studies is the difference between macro- and microtechnologies introduced by the anthropologists H. Russell Bernard and Pertti Pelto. Macrotechnologies involve "large-scale environmental transformations" such as dams, factories, new cities, etc., whereas the microtechnologies are "machines and equipment that individuals or organizations can acquire in the open commercial market" (Bernard and Pelto 1987:6–7). Macrotechnological projects are generally controlled by corporations, governments, banks, international agencies, or other nonlocal institutions, and those institutions frequently relocate indigenous populations and take over their land and water resources. In contrast, microtechnologies usually reach indigenous communities through markets, missionaries, or aid agencies. As a result, it is more likely that local communities will have greater control over microtechnologies.

The distinction between macro and micro technologies therefore tends to coincide with nonlocal versus local control. However, as useful as the distinction can be in some situations, it should not be misused so that it becomes a form of technological determinism. There are important exceptions to the pattern, and indeed there are some cases in which macrotechnologies are associated with local control and microtechnologies play a key role in the wholesale destruction of a native people's culture. Furthermore, the very question of local control is complicated, for it can range from centralization in an authoritarian structure built around a few corrupt, older, male leaders from one region of the Native society or it can be decentralized across gender, geography, age, and other social divisions within the Native society. Likewise, even if the new technologies are accompanied by some version of local control, they may change the social structure and power relations within the indigenous society.

Thus, a second, and equally important, step toward a framework for analyzing the impact of cosmopolitan technology on indigenous peoples is to situate technological relations as part of a larger web of intercultural power relationships. In general, as national societies have expanded their frontiers in colonial and neocolonial ventures, economic and political motives have been primary. Cases such as the colonization of New England, where religious minority groups were attempting to build new societies, are relatively rare. In contrast, the European colonization of Africa, Asia, and the Americas is largely a story of geopolitical rivalry for control over trade routes, slaves, and natural resources. Clearly, the world is more complicated today: former colonies turned nation-states engage in their own internal colonialism; East Asian and other nation-states now compete economically with former European colonizers; economic neocolonialism has replaced the political colonialism of European empires; and the sweat shops of poor countries have replaced the slave ships of the colonial era. Still, political-economic relationships remain as crucial to understanding the encounter between national and Native societies today as they did in previous centuries.

Finally, while the macro-/microtechnology distinction, the question of local control, and the framework of global political economy are necessary starting points, they do not provide a sufficient framework for a discussion of technology, Native peoples, and local power. Theorists of monolithic structural forces can reduce Native peoples to passive victims of colonizing forces, and in the process these theorists unintentionally contribute to the very colonizing forces that they purport to criticize. In most cases Native peoples are actively involved in redefining, reconstructing, and resisting cosmopolitan technologies, and they often incorporate cosmopolitan technologies into their own resistance struggles. Thus, the study of cosmopolitan technologies and Native peoples needs to be balanced between a framework that examines the social impact of those technologies and one that looks at how the technologies are reconstructed and resisted. However, before going on to the question of reconstruction and resistance, it is useful to flesh out some of the findings of the first wave of research that examined the question of the social impact of cosmopolitan technologies on indigenous societies.

Microtechnologies

As Bernard and Pelto recognized, microtechnologies are by no means always associated with a locally controlled process of benevolent technological change. Microtechnologies can lead to "culturecide," the collapse of the fundamentals of a culture: its religion and cosmology, its social structure, and its system of decision making and conflict resolution. Perhaps the most well-known example is the case of "Steel Axes for Stone Age Australians," described by the anthropologist Lauriston Sharp (1952). His essay is an example of the kind of cultural breakdown that can occur as cosmopolitan microtechnologies diffuse into Native societies. Sharp thought such breakdowns might be mitigated or even avoided if well-trained anthropologists were placed between the local tribal communities and the forces of development.

Sharp's article details the effects of steel axes on the Yir-Yiront, a hunter/gatherer/fisher society of the Cape York Peninsula of Queensland, Australia. By 1915 an Anglican mission had been established at a distance of about three-days march from the main part of the Yir Yiront territory, but in the 1930s the Yir Yiront were still living in relative isolation and without much exchange with or influence from the outside world. Over the years, however, the missionaries began to give certain goods to the Yir Yiront, and nothing was in higher demand than the steel axe, which was used for construction, cooking, hunting, gathering, and other activities.

Prior to the missionaries' introduction of steel axes, Yir Yiront men were the "owners" of stone axes. They obtained them through a series of complicated trading relations with men of other tribes. Although women and children could use an axe, they had to borrow it from a man and return it promptly. The borrowing rules followed and reinforced the kinship system: a woman borrowed from her husband, or from her older brother or father when it was not possible to borrow from her husband. Likewise, children and younger men who needed an axe borrowed it from their older brother or father. The borrowing system therefore confirmed a kinship system that involved a social hierarchy of male over female, older over younger. Furthermore, during rituals only members of one clan—the Sunlit Cloud Iguana people—could use the axe, and as a result the stone axe played a supporting role in the totemic structures of the whole religious and ritual system.

The older men, however, had a better collective memory of the white man's harshness, and consequently they were more suspicious of the missionaries and the white man than were the younger men. Thus, the older men were less willing to interact with the missionaries to get the steel axes. The missionaries also tended to hand out steel axes indiscriminately, even to women and children. As a result, conventional borrowing relations by gender and age broke down, as did trading relations with other tribes, and older men were forced to borrow axes from women or younger men.

One might ask at this point if it is "good" to introduce new technologies that may undermine existing social structures when, from a Western perspective, those structures could be categorized as sexist or ageist. After all, why should women and younger men be restricted from access to the axe? Maybe it *was* good to have a cultural revolution within the Yir Yiront, so that the powerful older men were forced to ask for steel axes from younger men and women. Certainly the argument could be made that the breakdown of age and gender hierarchies in Yir Yiront society was good because such social forms were unjust, and because equality is better. But who gets to define what constitutes justice and a better social order? Is equality always better than hierarchy? How does one trade off a sense of universal human rights with the colonialist implications of interfering in another society?

Certainly, I tend toward toward the hands-off side of the spectrum. I think the only just answer to this moral dilemma can be to let the Yir Yiront themselves decide what they want from cosmopolitan societies and how they want to interact with those societies. However, this ideal solution rarely is implemented, because Native peoples generally have no voice in the national society, and they are relatively powerless when compared to other interests that want to colonize them and their lands. Furthermore, even the hands-off solution still begs the question of who among the Yir Yiront gets to decide and how they should decide. Is the solution just if the male elders hold a council and make a decision? It may be, in terms of the definitions of justice and proper decision making among older Native men, but will the younger men and women agree? My own answer is that if disempowered groups within the Native society want change, and if I agree with the changes they want, then I may feel justified in helping them, because there is more than one definition of justice within the society. However, in the case of the Yir Yiront, Sharp does not tell us what the women and younger men wanted. They may have wanted steel axes, but they may as well have wanted to avoid all the implications of cultural collapse that accompanied their introduction.

The overall impact of the introduction of the steel axes and contact with the white man is probably something that all Yir Yiront would agree was undesirable. Yir Yiront ended up going to work for missionaries and cattle ranchers, and in the process they became enmeshed in new hierarchies and inequalities that were probably worse than those of their own society. More important, the introduction of the steel axe contributed to the demise of Yir Yiront cosmology. The cosmology was closely linked to the totemic clan system. One of the clans was called the Head-to-the-East Corpse clan. The clan was associated with death, the color white, and ghosts. When the whites entered the scene, they and their products—including the steel axe—were naturally associated with that clan. However, the traditional stone axe was a totem of the Sunlit Cloud Iguana clan mentioned earlier. Sharp continues:

> Moreover, the steel axe, like most European goods, has no distinc-
> tive origin myth, nor are mythical ancestors associated with it. Can
> anyone, sitting in the shade of a *ti* tree one afternoon, create a myth
> to resolve this confusion? No one has, and the horrid suspicion
> arises as to the authenticity of the origin myths, which failed to
> take into account this vast new universe of the white man. The
> steel axe, shifting hopelessly between one clan and the other, is not
> only replacing the stone axe physically, but is hacking at the sup-
> ports of the entire cultural system. (Sharp 1952:22)

A collapse of cosmology accompanies the collapse of the social order.
Sharp portrays the next stage of the culture collapse that he has seen
with other, more assimilated Yir Yiront groups: the demoralized Natives
who become assimilated into the cattle stations or frontier towns.

Sharp's study of steel axes is instructive, for it shows how even apparently innocuous technological innovations such as steel axes can entirely transform a Native society. However, other studies of microtechnologies and indigenous peoples suggest that the results are not always
as disastrous. The well-known case of the acquisition of snowmobiles
among the Eskimos of northern Canada and the Lapps of Norway provides a counterexample in which microtechnologies had a relatively beneficial impact (Pelto 1973, Pelto and Müller-Wille 1987). Unlike the Yir
Yiront, the Eskimos and Lapps had lived in contact with the national
populations for many years, and they were more or less in control of
the introduction of the new technology. As a result, the introduction of
the snowmobile was a relatively benign process. In general, the Lapps
and Eskimos decided that the advantages of increased mobility and
decreased animal care outweighed the disadvantages of increased cost
and increased risk due to breakdown. Once a few members of a community successfully experimented with the new technology, others usually
followed. As the anthropologists Pertti Pelto and Ludger Müller-Wille
note, "It comes as something of a shock to our romanticism to learn that
these people have been relatively quick to kill or sell their dogs and rein-
deer in order to buy snowmobiles" (231).

Because Pelto and Müller-Wille collected information on a comparative basis, they were able to extend social impact studies by studying
how acquisition and impact are adapted to the local environment, an
issue raised by Steward (1955) in his theories of cultural evolution. For
example, in the southernmost of the regions studied in Canada, Belcher

Island in Hudson Bay, dogsleds provided the advantage of better mobility over the rough terrain of the island, and the sleds were also useful over a longer period of the year, especially when ice was beginning to melt (Pelto and Müller-Wille 226–27). Pelto and Müller-Wille also extended the social impact studies by examining how social equality played a role in the differential impact of the innovation (238–39). In areas where there were already cleavages between haves and have-nots, the snowmobiles appear to have made the gap worse. However, in areas where everyone has been able to afford a snowmobile, there has been a relatively small impact on social inequality. Furthermore, in many areas the snowmobiles have made it easier for people to visit one another, and consequently they have led to an increased intensity and frequency of social relations.

Yet another dimension to the question of microtechnologies and indigenous peoples is brought up by the anthropologist Lila Abu-Lughod (1989), who has studied television and cassettes among the Bedouin of western Egypt. As with the Arctic groups, the Bedouin have had a long-term relationship with the surrounding national societies, and they have had relatively good control over the introduction of microtechnologies. Thus, one would expect the social impact to be relatively benign, and indeed that is what she describes. In many cases watching television and listening to cassette recordings becomes just one more part of daily family activity. As in the case of the steel axes, Abu-Lughod shows how the new technologies have broken down gender and age hierarchies, but in a more benign way. Men and women, old and young, tend to watch programs together, and television has provided a welcome window on the outside world for sequestered women.

Abu-Lughod adds to the literature on the impact of microtechnologies by questioning the widely held idea that the new media technologies will lead to cultural homogenization. She notes how families use the songs, poetry, and television programs as occasions for commentary, and they also approach the media selectively. (I can confirm the point by my own experience of watching the television program "Dallas" in Brazil: my friends had a much more favorable reaction to J. R. than I had seen in the United States.) Songs, poetry, and television programs are therefore good examples of what has been called "interpretive flexibility": the idea that new technologies are often used and interpreted in ways quite at odds with their authors' or producers' intentions.

Furthermore, Abu-Lughod found that the Bedouin were recording their own poetry and songs on cassettes, and one Bedouin singer had even achieved something of a hit status in the national society. The national society's interest in the Bedouin singer helped promote pride in Bedouin culture, but at the same time Abu-Lughod's informants also told her that they preferred another, lesser known Bedouin singer who had refused an invitation to record in a Cairo studio and who was also less "Egyptianized" than the first singer. Thus, microtechnologies can contribute to a reverse social impact of counteracculturation: the indigenous society can affect and shape the national society.

Nevertheless, even the process of counteracculturation is not pure. The Bedouin singer became a star in the broader Egyptian national culture, but he also went electric. Thus, as aspects of the local culture are absorbed into the national culture, they are also transformed. In the process new hybrid cultures and cultural forms emerge. Rather than think of science and technology as making the world more homogeneous, it seems to make more sense to think of science and technology as helping to create new hybrid cultures, a process sometimes referred to as ethnogenesis.

Macrotechnologies and the Limits of Interpretive Flexibility

Today, the chaos that Sharp described for the Yir Yiront is idyllic in comparison with the impact of large-scale macrotechnologies and development schemes in many parts of the world. The steel axes of yesterday have given way to chain saws, not to mention hydroelectric dams, mines, and roads. Furthermore, these macrotechnologies have increasingly come to dominate the interface between national and Native societies. Controlled by alliances of international and national capital, the so-called development projects all bring with them cosmopolitan technologies that destroy not only the economies and cultures of tribal and peasant populations but also the very land and environment itself.

In cases where there is no mediation by the state, such as in parts of the Amazon where development is largely controlled by ranchers and other local elites, Native peoples who get in the way of development schemes are likely to be forcibly moved out of the way or murdered. Even microtechnologies can take on an ugly face, as in the case of a group of Nambiquara Indians whose story is told by the anthropologist David Price in his moving account, *Before the Bulldozer* (1989). They woke up in the middle of the night to the sound of bulldozers invading their village to plow their huts into the ground, and the lucky ones were those who survived to tell the story. Cases of genocide have been reported from the Amazon, Central America, and Southeast Asia, to name only a few recent examples. The genocide of Native peoples throughout the world, and particularly in the New World, has achieved epic proportions that have sometimes been compared to the Holocaust of European Jews during World War II.

Even when governments and other organizations mediate the development process, the basic principle is still the same: to get the Native peoples out of the way of progress. However, their kinder, gentler solution is often forced resettlement, usually to smaller territories and often to areas with radically different ecologies. Even when the gross crises of sedentarization, disease, alcoholism, dependency, and socioeconomic collapse can be overcome, nothing can compensate for the loss of land. For Native peoples land is life; it is the text upon which the sacred cosmology and history of the people is written. To take away their land is to destroy their culture.

There are so many cases of macrotechnologies, relocation, and genocide that it is only possible to give a sketch of the range of issues involved. Perhaps no case is more memorable than the resettlement of Marshall Islanders of the American "Trust" Territories for the purposes of nuclear testing during the first years of the cold war. The American government took advantage of the local people's naïveté to convince them that if they gave up their island atoll for nuclear testing, then nuclear energy could be safely harnessed to benefit the human race and end all wars. The people of Rongelap were not removed, even though they were directly in the line of fallout and became human guinea pigs for the experiments. The people of Bikini Island were slightly more fortunate: they were moved three times to islands that were ecologically very different from their original atoll. As a result of the resettlement, their economy collapsed into dependency on the American government. Today, the Bikini Islanders are negotiating a return to their home under the cloud of uncertainty about the safety of their topsoil, flora, and fauna. Furthermore, the prospect of global warming and higher ocean levels raises the specter of the ultimate degradation of their territories: inundation.

At the time of nuclear testing, the Marshall Islanders, like many Americans, believed the American government's rationale for destroying their lands in order to test nuclear weapons. Today, however, Native peoples (like local communities in national societies) are generally much more sophisticated about cosmopolitan politics, and they are less likely to surrender land or resources without a fight or serious concessions. In some cases Native peoples are able to stop development projects, and in others they achieve a modicum of local control, at least over the revenue generated by the projects. For example, when the American government proposed its oil drilling plan for Alaska's North Slope, the Inupiat set up their own local government, the North Slope Borough, which had the power to levy property taxes on the oil projects. They used the revenues to launch a half-billion dollar capital improvement program that employed over half the Inupiat population building schools, roads, housing, and other public works. Household incomes went up, income inequality did not worsen, and the Inupiat even began college-level education in their language. Still, the results have been mixed: alcoholism and other social problems continue to rage, and roughly "equal proportions (20 percent) of [the] residents perceived the overall effects of oil development as good, bad, and mixed" (Kruse, Kleinfeld, and Travis 1984:332). As one Inupiat said, "Materially we're better off, [but] the culture is being lost faster" (332).

The Inupiat case shows that even with large-scale macrotechnological development projects, some indigenous peoples may be able to earn economic concessions even if they do not obtain enough control to manage or stop the projects. However, it does not seem appropriate to use examples such as the Inupiat to argue that even these macrotechnological projects are subject to "interpretive flexibility." As the political theorist Langdon Winner (1986) has argued, artifacts of all sorts—microtechnologies, macrotechnologies, buildings, roads, etc.—have politics. I would add that big artifacts in the form of macrolevel development projects carry with them big political implications, especially for the relatively small-scale indigenous societies that are usually pushed out of their way. Thus, while in some cases it is possible to speak of the interpretive flexibility of artifacts, particularly in the case of microtechnologies in situations of long-term contact and local control, those discussions should not ignore the large-scale structural constraints that limit interpretations and flexibility. Particularly in the case of macrotechnologies of development and Native peoples, to say that technologies are texts that can be interpreted flexibly ignores the political and economic structures that are rapidly destroying the lands and cultures of

Native peoples and the environment of us all. As Winner has noted, "Interpretive flexibility soon becomes moral and political indifference" (1993b:372).

Development, Technology, and Resistance

For an alternative to the approach to technology as a text that can be interpreted flexibly, Winner suggests an analysis of the larger political and economic structures in which such technologies are embedded. Analyses based on concepts such as class interests, multinational capital flows, and neocolonial state relationships provide a necessary backdrop for the process of strategizing how to fight back. Thus, the approach I adopt here assumes that most macrotechnological development projects are occasions in which various alliances of finance, government, and industry expropriate Native peoples' lands in their search for natural resources and profits. What I wish to focus on, however, is not the overwhelming economic forces that are driving these projects. Those forces can only be changed by the international struggle for greater economic equality, political democracy, and environmental consciousness in the cosmopolitan societies. Instead, I will focus on the specific resistance strategies that Native peoples use against the powerful forces that are destroying their lands and cultures. In doing so I move into the second level of analysis of technology and indigenous people, from social impact studies to research on resistance strategies and struggles.

Types of Resistance Strategies

Native peoples do not have time to wait for long-term struggles to reach their positive fruition. Throughout the world they are engaged in resistance struggles. In general resistance struggles tend to begin with negotiations through official channels. In some cases indigenous organizations have been able to achieve fairly successful concessions through direct work with regional or national governments. The success of that strategy is highly sensitive to cultural variations in national political systems. Successful negotiations through official channels tends to occur in democratic countries during periods when they are controlled by fairly progressive, or at least not anti-Native, governments. Cases of that type include some negotiations between Native peoples in the polar regions and the governments of the United States and Canada, the Sami in the Scandinavian countries, the Ainu in Japan, and the Aborigines in Australia during the years of the Whitlam government. Not all direct negotiations have been successful, and in some cases direct negotiations were successful after Native peoples entered a second level of resistance: nonviolent action.

As a resistance strategy nonviolent action is culture-dependent. My view is not universally held, and some peace researchers such as Gene Sharp have shown how nonviolent action can be successful even under conditions of extremely repressive authoritarian regimes. In *The Politics of Nonviolent Action* (1973) Sharp divides nonviolent resistance strategies into three main groups: protest and persuasion, noncooperation, and intervention. For indigenous peoples nonviolent action has worked fairly well in democratic countries, especially where it is possible to get media attention and to sway public opinion. In India, for example, where there is a longstanding tradition of nonviolent action, members of the Chipko successfully used those methods (including forming human rings around trees) to prevent a government-sponsored logging concession, and the movement, largely made up of women, spread to other groups. In Sweden the Sami chained themselves to bulldozers to prevent the construction of a hydroelectric dam, and their actions caused an immediate swell of sympathy among a public that prides itself on leadership in human rights and welfare. In Australia aboriginal protests gained the support of labor unions, which in turn struck to prevent a development project. In both Australia and New Zealand Native peoples have won concessions by setting up tent embassies in the national capital, a strategy similar to the famous Trail of Broken Treaties March on Washington, D.C., in 1972. In short, marches, protests, strikes, roadblocks, demonstrations, and any of the other dozens of methods of nonviolent action catalogued by Sharp can often be a first step toward building coalitions with sympathetic groups in the national society as well as with other Native groups.

Another form of nonviolent action involves entering electoral politics in countries where that option is available. In several cases, notably India, Native peoples have stood for election or even formed political parties. Whereas in some countries the strategy is successful, several Native American nations in the United States reject the strategy because it implicitly means recognizing their status as American citizens rather than as illegally colonized peoples whose lands have been stolen. However, not all Native Americans hold this opinion; for example, a Native American was elected to the U.S. Congress in the early 1990s and has pledged to stop the termination of Native lands.

A third level of resistance involves violence. Advocates of nonviolent action such as Sharp believe that violent resistance is ultimately self-defeating, but my experience as a Latin Americanist and an anthropologist makes me cautious about making universal claims for strategies and tactics in the resistance process. In some cases violent action has been limited to acts of sabotage. Frequently, sabotage is directed toward the machinery of development, such as bulldozers, but the target varies by site and development project. In Panama the Kuna Indians burned down hotels that were built on their lands without their permission and over their protests. Violent indigenous resistance has also gone beyond sabotage to the formation of guerrilla movements. Guerrilla movements tend to emerge where official channels and nonviolent options have been exhausted or where nonviolent forms of protest have met with ruthless repression. In many cases guerrilla movements are part of national liberation struggles, and in some cases indigenous guerrilla groups are linked to other liberation struggles. In guerrilla situations civilian populations are frequently forced to flee, thus causing huge flows of refugees, which has been classified as a form of nonviolent action.

To summarize, resistance strategies take different forms according to the political climate of the national culture and the phase or level of resistance. In relatively democratic countries resistance strategies sometimes do not need to go beyond official negotiations and nonviolent protest. In relatively authoritarian countries those avenues may be closed, and nonviolent action may lead to brutal repression. In situations of uncontrolled state or local fascism, the only viable option may be guerrilla activity or simply giving up and fleeing the area. To insist on nonviolent action as a matter of principle in all circumstances may in effect result in genocide; in other cases violent action may lead to even greater repression and genocide. In other words, strategies and tactics cannot be theorized universally; they must be determined locally.

The success or failure of indigenous resistance movements is not wholly determined by the type of national and international environment in which they are operating. What the peace researcher Johan Galtung has described in his essay "A Structural Theory of Imperialism"

TABLE 8.1
Technologies of Destruction and Resistance

DESTRUCTION	RESISTANCE
Extractive Technologies mines, dams, military testing sites, logging	Technologies of Infrastructure airplanes, boats, videocassettes, shortwave radios
Technologies of Settlement and Colonization roads, airstrips, pesticides, and farming technologies	Indigenous Media video, film, musical recordings, indigenous material culture
Technologies of Domination and Surveillance satellites, airplanes, weapons	Technologies of Guerrilla Warfare medical, infrastructure, weapons

(1980) as the feudal interaction structure of the world system's peripheral regions can also affect success. In other words, resistance movements can become fragmented where there are great geographical hurdles, where various indigenous groups lack a common language, and where indigenous nations are divided by national frontiers. Furthermore, in countries such as the United States resistance is sometimes complicated by a layer of corrupt tribal leaders whom the government and development interests have bought off in order to give a veneer of legality to the development projects. In those cases official channels are doubly blocked by a conspiracy of tribal leadership and development interests, and underground, guerrillalike activity may become necessary.

Technologies of Destruction

The problems of indigenous people are political, military, and economic ones, but the politics of technology also play a role in shaping the texture of the conflicts. To develop an understanding of the role of technology in the destruction and expropriation of Native people's lands, and in their resistance struggles, it is necessary to develop a new kind of vocabulary that goes beyond dichotomies such as macro- and microtechnologies. A better way of thinking about technology in the context of Native people's land and human rights struggles is to think in terms of technologies of destruction and resistance (see table 8.1.).

Not all technologies employed by national societies and international organizations are technologies of destruction. There are numerous cases in which hospitals, schools, and other projects do indeed seem to meet with the approval of Native peoples and can be considered technologies of empowerment. However, my concern here is with establishing a typology of technologies of destruction. I divide the technologies of destruction into three groups: technologies of extraction, settlement, and military domination.

Technologies of extraction include macrotechnological development projects such as hydroelectric dams, logging and mining operations, and military testing. Those projects extract energy, materials, or knowledge by destroying the landscape. Extractive technologies are not necessarily associated with huge influxes of population, and as a result there is at least a possibility that the environmental damage and land loss can be contained. Extractive technologies involve the displacement of local Native peoples, but the extent of the displacement depends in part on the scale of the project and other technical issues. Military testing (especially if it is nuclear weapons testing) usually involves such widespread and dramatic damage to the environment that the Native society is forced into radical relocation and resettlement. In contrast, with logging, hydroelectric, and mining operations Native peoples have a better chance of continuing to exist beyond the edges of the operations.

A number of technical issues have dramatic environmental implications. For example, the damage of logging can be limited if the forest is not clearcut, and likewise the damage of hydroelectric projects can be limited if big dams are replaced by a series of small hydroelectric projects. (The smaller dams are generally more efficient both economically and electrically, and they produce less environmental damage.) In cases such as the James Bay hydroelectric projects in northern Quebec, the failure to remove trees and flora has resulted in high levels of mercury pollution, which in turn has made traditional fishing impossible in some areas. Likewise, extractive projects such as the Carajas project in Brazil originally involved deforestation for charcoal production, rather than using coke from outside the Amazon. Thus, a number of ostensibly technical decisions can affect the extent of the destruction caused by extractive technologies.

Extractive technologies are not necessarily accompanied by settlement, but they may be, especially if roads are opened up. Roads are the primary settlement technology, and again there are technical issues among the settlement technologies that have enormous environmental and political implications. For example, railroads and airstrips may cause less damage than highways, because highways open up an area to massive settlement and cannot be as readily monitored or controlled. Even within the highly destructive settlement technology of roads, there are technical issues that may mitigate their harm. For example, if roads are built away from population centers of Native peoples, their damage may be restricted. Likewise, if roads are built out of materials from other regions of the country and with nonlocal contractors, there may be a better chance that less ecological damage is done to the local region and construction crews will go back to where they came from after they have finished their work.

Settlement technologies also include technologies of infrastructure and of farming. Automobiles, farm equipment, radios, and airplanes may be considered microtechnologies, but that does not make them any less damaging. Indeed, settlement technologies are arguably more damaging than extractive technologies, because they are the technological basis of new populations and modes of production that are less likely to coexist peacefully with Native groups. Whereas mines, dams, and limited logging operations may leave some maneuvering room for Native peoples, settlement technologies usually imply larger levels of displacement. In the case of forest-dwelling peoples, colonization is likely to result in massive burning to clear the land for farms and ranches. The new settlement technologies also bring with them their own forms of pollution from pesticides, increased risk of disease due to enhanced population contact, and a high likelihood of violent confrontation over land rights.

Finally, technologies of domination and surveillance refer to the military and police technologies that enforce the new order. The military technologies include not only control over superior weapons but also control over access to the means of transportation and communication, including airspace. In some cases, the technologies of domination have become part of genocidal projects aimed at eliminating or enslaving Native peoples. Highland Guatemala, West Papua, the Chittagong hills of Bangladesh, and the villages of the Aché and Guayakí of Paraguay are among the most notorious cases in which technologies of domination have become the instruments of genocide.

A typology of technologies of destruction may help sympathetic groups in the national societies to interrogate development projects that they have little chance of stopping, by pointing to technical issues that might be changed in order to mitigate the destruction. However, more effective strategies will have to involve directly helping Native peoples in their resistance struggles to stop the incursions in the first place.

Technologies of Resistance

In Native resistance struggles that opt for violent action, military technologies emerge as the most important. Foreign governments can help by cutting military supplies to the offending national government, demanding an end to genocidal policies, offering medical aid to indigenous populations, and providing relief for refugees. Although it may be illegal for pro-Native activists to fund directly Native guerrilla movements, activists may still provide general funds as well as publicity and lobbying in the national governments.

In the official and nonviolent types of resistance struggles, technologies of resistance are, principally, technologies of infrastructure. For example, in Brazil the Kayapó have their own "air force," which they use to patrol their lands for squatters and other invaders. They also document statements by government representatives with video, so that they can later show how the government representatives lied or broke promises. Shortwave radios allow for communication among villages, and airplanes allow tribal leaders to travel among themselves and meet with each other. Boats and buses allow Native peoples to move greater numbers of people. The anthropologist William Fisher, who did fieldwork among the Kayapó, further explained in a letter to me that these technologies of infrastructure—planes, radios, and boats—were originally used by the government and development companies to dominate and control the Kayapó, but subsequently the Kayapó have either purchased the technologies or pressured the government or development companies into letting them use the technologies.

Technologies of infrastructure allow Native peoples to move information and people with greater liberty. That in turn makes it possible to expand and strengthen coalitions and also to amass larger numbers of people for public demonstrations. However, as Fisher notes, the idea of coalition politics among Native peoples is not always the same as in urban, Western settings. Indigenous groups draw on kinship links, age groups, ritual obligations, and other cultural resources for mobilizing people to joint action.

In addition to the infrastructure technologies, there is a relatively new form of technology that seems to be increasingly effective in nonviolent protests, particularly those that capture international public attention: indigenous uses of media. In the past anthropologists sometimes made ethnographic films, but in recent years indigenous peoples have been producing their own video, film, and other electronic media. Some groups—notably the Kayapó of Brazil, the Inuit of Canada, and the Aborigines of Australia—have become very sophisticated at the use of media in their efforts to maintain their culture and to resist outside interference.

Indigenous media and the use of international media is now gaining recognition as a resistance technology of great potential. The film *The Kayapó: Out of the Forest,* directed by Michael Beckham with the supervision of the anthropologist Terence Turner, documents such use for a demonstration of indigenous Amazonian peoples who opposed a huge hydroelectric scheme on the Xingu River that the government had proposed. The Kayapó and their supporters at the Altamira demonstration were very successful at generating media attention for their cause. Part of the success resulted from the support of the rock star Sting, whose participation helped attract international media attention. Media stars have increasingly realized their power as lightening rods that can bring media attention to causes that would otherwise go unreported.

Although the appearance of an international star is helpful for gaining a toehold in the international media spotlight, perhaps less well understood is the role of traditional material culture (for example, clothing and tools) in the mediated productions of indigenous peoples. The Kayapó-led demonstration at Altamira involved an odd mixture of nonviolent public protest, staged ritual warfare, and highly sophisticated international media manipulation. To some extent, the mixture was due to the different levels of understanding and sophistication among the coalition of forest peoples involved. The leaders and the anthropologists clearly understood the importance of the media, but some Native peoples used lip plugs and gave speeches in traditional style, or they brandished clubs and spears and staged war dances. Some of the Kayapó also appeared dressed in traditional garb and at the same time carried video cameras to film the event. The juxtaposition of video cameras and lip plugs, media event and war dances, emerged naturally from the different levels of understanding and strategic use of those levels for the purposes of the demonstration. The low technology of the lip plugs and war clubs fed into the high technology of the TV and video cameras to convey an image that would play on the public's romantic sensitivities for the Native peoples.

Fisher (1994) has examined the Altamira protest in the context of the history of Kayapó relations with the national society. His analysis demonstrates how the success of the Kayapó's use of Native adornments and the media was a highly contingent strategy used in a long-term struggle for self-determination. He demonstrates that there were various phases of relations between the Kayapó and the outside world that were shaped by the regional political economy. More specifically, the external threat changed from Brazil nut and rubber harvesting in the 1940s and 1950s, to the national government's road construction projects of the 1960s and 1970s, to large-scale mining projects (and the Xingu dam project) supported by international capital in the 1980s. The Kayapó have been very savvy about responding to external colonization schemes, and they have tailored their resistance strategies to the specific economic or technical forces of outside domination. Thus, as Fisher argues, the Kayapó's media-saturated protest at Altamira, and their self-conscious utilization of urban environmental sympathy, were part of an ever changing and flexible struggle of resistance.

Fisher contextualizes the Kayapó protest at Altamira by situating it historically, and I would add to his analysis that such media-oriented strategies might also be examined comparatively in order to find other patterns. The Kinzua Dam case of the United States in the 1950s provides an interesting point of comparison. Like the Kayapó the Native Americans were organized, but unlike the Kayapó they were not successful. The Kinzua Dam was built on the Allegheny River in Pennsylvania, and it flooded into Seneca reservation territory in the state of New York in clear violation of treaties. The Eisenhower administration followed a policy of termination and assimilation of Native American tribes, and Washington politicians supported the Army Corps of Engineers' project because it served the interests of Pittsburgh industrialists and voters. Prior to the construction of the dam, the Seneca people organized a letter-writing campaign, garnered the support of numerous intellectuals and even New York governor Averell Harriman, developed an alternative plan under the advice of the former TVA chair and civil engineer Arthur Morgan, issued numerous protests from Native conferences, and went to court. President Cornelius Seneca even went on the nationwide television program "To Tell the Truth," dressed in full Indian regalia, to win public support and publicity. For the 1950s it was a very sophisticated protest movement, but it failed to capture public attention and support. Probably at that time direct protest and confrontation at the dam site would have been the only tactic to capture public attention. Indeed, the struggle to stop the Kinzua Dam project, and the fight to stop the loss of other tribal rights and territories under the power projects of New York state's public works mogul Robert Moses, contributed greatly to the position of the Iroquois as leaders of the red power movement.

Let me return the focus, however, back to the question of technology, here taken broadly to include the role of Native material culture in the resistance struggle. The use of Native clothing, weapons, adornments, and other material objects may provide one means onto the stage of mass media, although it could also provoke a backlash among viewers if the usage were to be seen as manipulative. The symbols need to be thought through as part of a locally based set of tactics. For example, to protest New York State's plans to buy energy from Hydro-Quebec, the Cree and Inuit of northern Quebec paddled down through upstate New York and the Hudson River. As the STS researcher Kenneth Croes (1994) explains, they paddled on an *odeyak,* a mixture of a canoe (*ode* from the river-using Cree word for canoe) and a kayak (from the seafaring Inuit word). The use of the canoe-kayak was not disingenuous; it was an appropriate symbol that bridged the common interests represented by an artifact that Natives use for food gathering and transportation and that middle-class viewers of the national society could recognize as an instrument of recreation. The canoe-kayak reminded viewers of one aspect of the national material culture that is derived from Native culture, and it revealed a point of common cause: interest in environmental preservation. Croes reports, "Through this event the Cree were successful in galvanizing the support of the U.S. environmental community." So far, the New York State government has refused a second contract with Hydro-Quebec, although the lack of projected need for additional electricity is probably a more important consideration than the environmental impact of the project or the rights of the Cree and Inuit.

I do not know what to call the lip plugs, feather bonnets, war clubs, war paint, odeyaks, and other examples of indigenous material culture as they are transformed into symbols through television and the media. The term *cyborgs* comes to mind. Cyborgs bridge human and machine, just as lip plugs, war paint, and nose and earrings bridge the human with the material world of things. Yet, in the gaze of the television camera, the low-tech cyborgs become something else: symbols of a more natural way of life that provides an answer and alternative to the high-tech scientific concerns of ozone depletion, global warming, and ecological sustainability. As the media and public leaders of the 1990s recast the Armageddon script from the threat of nuclear war to global ecocide, Native peoples achieved new recognition as custodians of a message of ecological sanity and sustainability. However, if Native peoples have achieved a new status as the first ecologists, they also run the risk of becoming valued only as new versions of the romantic noble savages that each generation constructs as an answer to what it finds lacking in itself. As Fisher points out, the fundamental issue is not what Native peoples can offer national societies, but instead how national societies can be prodded into recognizing and honoring the rights of Native peoples to self-determination.

Although I have focused on indigenous media in this section as an important new resource among the technologies of resistance, it may not always be suitable to the specific context and problem. In some cases Native peoples have been forced into more direct forms of confrontation to defend their land. In cases where they themselves face direct military suppression, they may be left with no other alternative than to flee the area as refugees or to arm themselves and fight back.

Grassroots Development as Resistance

In a discussion of the history of the indigenous rights organization Cultural Survival, the anthropologist Jason Clay wrote about a change the organization gradually underwent: "We realized that advocacy, publicity, and press releases alone wouldn't save people, that most of our efforts and money should go to projects that indigenous people themselves design and run" (1992:31). In a similar way the concept of resistance and the relationship between technology and resistance needs to be expanded to incorporate what the Clay and his colleagues call "grassroots development." Here, the word *development* is used in a less sinister sense; it refers to projects not only for the benefit of indigenous peoples but also controlled by them. Thus, it is not enough to work to stop dams, mines, farmers, and ranchers; resistance must be proactive as well as reactive. Indigenous peoples need to have the economic and infrastructural abilities to be in a strong position to fight back when the dams, mines, ranchers, and other agents of destruction arrive.

Why Development Projects Often Fail

Most development projects today are designed in a government office, and frequently they become infused with governmental and industrial interests. Some anthropologists and other social scientists have played a role in development projects of that type, but their role is usually to mitigate the negative effects of projects that have already gained momentum. The power relationship between national and international elites on the one side and local communities on the other side is often reproduced within development projects in the relationship between economists and natural scientists on one side and anthropologists and other social scientists on the other side. As the anthropologist Janice Sacherer points out, "[A]pplied anthropologists are often hired only after deep conflicts and difficulties have surfaced and the project's basic structure has crystalized" (1986:249). Like ethnic minorities in American businesses, who have frequently experienced the rule of being the last hired and first fired, anthropologists are often at the bottom of the project totem pole. Fitting with that position, they are seen as lone rangers, "tortoises," "soft," "nostalgic," and "fully replaceable" (Rhodes 1986: 33–38).

The anthropologist Bryan Pfaffenberger (1988) argues that even well-intentioned development projects may fail if the planners adopt a naive stance toward technology and its relationship to the local culture. He demonstrates this point in a study of irrigation projects developed by the Sri Lankan government to help peasants. The general problem with the irrigation projects was that people who occupied the top end of the project—the part closest to the water source—used up more than their fair share of the water, leaving those at the tail end with little water. The result was that top-enders grew wealthy and the tail-enders became poorer. These social effects seem to be an inevitable result of the technological facts of gravity-driven irrigation systems. Only a much more extensive system of field channels and automated delivery would result in a more egalitarian system.

However, Pfaffenberger argues that in traditional Sri Lankan society there are longstanding customs that regulate access to water and mitigate the wealth differentials of top-enders and tail-enders. Instead of setting up regulations that would support systems of reciprocity between top- and tail-enders, the Sri Lankan government set up an individualistic system of property ownership that allowed top-enders to accumulate wealth but did not encourage them to share it. Pfaffenberger argues that the failure of the development project was due to a narrow-minded approach to technology that ignores its part in a complex web of social relationships. His cautionary tale could well be generalized to many other development projects. I will consider here only one case study that demonstrates many of the issues found in other top-down development projects.

The Green Revolution

Similar problems plagued the thinking behind the much larger-scale and more influential series of development projects known as the Green Revolution. At first glance the prospect of the Green Revolution is appealing and straightforward: scientists will breed new plant varieties that produce more food than the older, natural varieties, and as a result world hunger will be reduced or even solved. No one will lose, and the vast numbers of the world's poor will be much better off. Farm size does not have to be any bigger or smaller, so it is possible to avoid all that ugly discussion about land reform and politics. It is the perfect technological fix.

However, technology is always embedded in political agendas and economic interests. The development researchers Robert S. Anderson and Barrie Morrison argue for an approach to the Green Revolution that situates it against the background of World World II and the end of British rule in southern Asia. India, Pakistan, Bangladesh, and Sri Lanka were all undergoing such rapid population growth that their populations doubled in size from 1950 to 1980. Those countries came out of World War II with weak central states that had a poor tax base. Japan had been defeated, but the cold war was raging. The rapid population growth, weak central states, and cold war climate contributed to growing communist movements in the rural areas of South Asia. Rural development seemed to be necessary.

The leaders of the new Asian states—together with their anticommunist supporters in North America, Australia, and Western Europe—could have chosen a democratic form of rural development. In other words, they could have opted for land reform: to limit land holdings, to create tenant protection legislation, to set up cooperative credit organizations, to develop government price stabilization programs, and to expand grain storage programs. However, the land reform strategy would probably not have strengthened the weak central states, whereas technological aid was more amenable to that goal. Anderson and Morrison argue that within the technological option the development bureaucrats had two choices: biochemical aid, such as fertilizers, pesticides, and new high-yield varieties (HYVs) of seeds; and mechanical aid, such as water pumps, tractors, drills, threshers, and so on. The state planners believed that the first solution, biochemical aid, was scale-neutral and land saving, whereas the second was not. In other words, under the first solution both large and small farmers could benefit, and because they used the new HYVs of rice and other crops, they would produce more food without needing much more land. Thus, within the technological option that increased their own power, the planners chose the more equitable alternative.

However, the Green Revolution turned out not to be as equitable as first thought. To begin, the scientific research behind the Green Revolution was not structured to meet local needs. The international research institutes for the different crop types did not decentralize their work into various countries or local communities, where research could better take local needs and ecologies into account. Furthermore, the new varieties of seeds emerged from laboratory research, not from local communities, where there were hundreds of varieties of nonstandard seeds that often matched the local ecology in ways that had never been studied.

Who benefited? The state bureaucrats benefited, because they controlled the distribution of the new HYVs. The elites of the Western countries benefited, because they entered into a partnership with third world countries that became partially dependent on Western funding, research, and technology to support the Green Revolution. In the case of rice, the international rice companies also benefited, because the standardization of rice varieties simplified their work of storing, milling, and shipping. Petroleum-producing companies and countries also benefited, because the fertilizers and pesticides that accompany the HYV crops are made from petroleum products. The central state governments therefore may have gained power with respect to their regional and local populations, but what they gained locally they lost globally by incurring deeper international dependency relationships.

Did the rural poor benefit? Yes, in some cases they did benefit, at least somewhat. Answering that question completely would require attention to the variations across the many regions of different countries. However, in general the Green Revolution turned out not to be at all as scale-neutral as originally touted. Richer peasants had more access to government services, which meant better access not only to the rice but also to fertilizers, pesticides, and other inputs to agriculture. The rich got richer and the poor got poorer. In some areas richer farmers expanded and displaced smaller ones, and the result has been polarization, violence, and bloodshed: the Green Revolution turned red. Such problems could only increase the pressure on indigenous lands as well as increase the possibilities for the very communist insurgencies that the planners hoped to stave off in the first place.

Although there is less agreement in the literature on the issue of gender, the Green Revolution also had a differential impact by gender. One effect was that because women and children—especially in poor areas of the third world—often suffer disproportionately in terms of nutritional status, it is likely that the increased income gap led to increases in the gender and age gap among those peasants who became poorer. Furthermore, there is also some evidence that even among those peasants who were better off, some of the traditional forms of patriarchy may have been perpetuated by other means. For example, men have tended to go into cash farming or wage labor jobs with higher pay, whereas women have continued to do subsistence and food-producing farming, or when they work for wages they do so at a much lower rate than men. Furthermore, ultrasound tests have increased in some Green Revolution areas, thus allowing a continuation of the practice of female infanticide via abortion. The correlation may not be causal, but it is also possible that the increased wealth has made the misogynist use of reproductive technologies more accessible.

An additional and unforeseen impact of the Green Revolution has been that excess food produced by HYVs does not necessarily end up feeding the poor. In Mexico, for example, HYV crops have been fed to animals to produce beef and cheese to sell to the urban, middle classes or to international markets. As a result, the Green Revolution in Mexico accompanied a change in the country's status to that of a net importer of food. Finally, as standardized HYVs have replaced indigenous crops throughout the world, the agricultural gene pool has been reduced. Standardization may have made it easier for multinational companies to sort, mill, and ship crops, but it has also made local crops more vulnerable to pestilence. A diverse crop base protects against disease because diseases do not always jump from one strain to another. Furthermore, local varieties are often better adapted to fluctuations in temperature and rainfall that could threaten more standardized HYVs.

The Green Revolution has therefore led to a new scientific problem: saving the world's genetic diversity of crop seeds. As universities, governments, and corporations talk about diversity issues in their new, increasingly multicultural societies, scientists now begin to worry about the diversity of the agricultural gene pool. The Green Revolution, with all its modernist assumptions of big-is-better and standardization, has given way to what we might call the Gene Devolution. Scientists scurry around the world to rescue rare varieties of seeds before they disappear entirely, and they store the precious seeds in central vaults. In the United States the vault has become known as the Fort Knox of genetic diversity.

The new solution smacks again of the technological fix. It involves the same ignorance of politics that characterized the Green Revolution, in which centralization of the same order created standardized varieties that were an important factor behind the very diversity problem that a new centralized institution is attempting to resolve. Centralization empowers those who control the centralized institution, and, to the extent that seed storage occurs in first world countries, neocolonial relationships are again being reproduced.

The technological fix of seed storage is also likely to fail for political and economic reasons. As the economic stakes grow, local governments are beginning to restrict scientific expeditions in search of seeds. After all, agricultural diversity is now part of a country's wealth, and it should not be given away freely. Local countries set up their own seed banks that centralize resources on a national rather than international scale. Furthermore, even the whole program of storing seeds suffers from long-term technical problems. Stored seeds must be taken out of storage and bred every few years, and with each new generation some of the original diversity is lost.

What alternatives are there? To begin, let us not forget the crucial step not taken at the beginning of the Green Revolution: agrarian reform. Decentralizing agricultural production means that more farmers are in control of planting decisions, and they are more likely to preserve genetic diversity than large agribusinesses or monocrop plantations. Furthermore, small farms with diverse crops mean a reduced need for pesticides. In the rain forests of Central and South America it is common for farmers to plant dozens of crops on one acre, mixing perennial and nonperennial crops. The ethnoknowledge on which intercropping is based represents a successful way to reduce pest infection and limit soil depletion. In India the Shri AMM Murugappa Chettiar Research Centre has adopted a similar strategy known as biodynamic gardening. Researchers from their center have worked with women to help them set up high-yield biodynamic gardening that relies on organic fertilizer (such as manure) and organic pesticides (plant-based). The spacing of the plants is intensive, but multiple crops are grown in one bed to limit pest infections and to complement nutrient needs.

At the other end of the process, consumer education is also important. For example, some scientists and entrepreneurs are already working to develop markets for unusual but hardy crops such as amaranth or blue corn. As consumers become aware of the importance of maintaining local variation in crops, they will become more willing to accept old foods in unstandardized forms as well as to develop a taste for new fruits and vegetables.

There are also technical solutions to the problem of developing higher yields that inscribe political relations other than increased dependency and centralization. As the biologist and STS researcher Meera Nanda (1991) points out, many of the seeds now produced are not hybrids and do not need to be replaced every year, and it is also possible to produce seeds that outperform standard varieties even when traditional agricultural methods are used (that is, without pesticides and chemical fertilizers). Thus, in theory the Green Revolution can be unpacked, reconstructed, and combined with traditional agricultural practices that do not strengthen dependency relations. Furthermore, given the tremendous flexibility of the new biotechnologies, it may be possible to recreate some of the genetic diversity that has been lost through standard varieties by engineering new, nonstandard varieties that are keyed to local ecologies and local knowledges.

Lessons from Grassroots Development Projects

The lessons of top-down development projects such as irrigation in Sri Lanka and the Green Revolution suggest that the appropriate alternative is grassroots development. In turn, the concept of grassroots development involves returning to the question of local control raised at the beginning of the chapter. It is now time to examine in more detail some of the complexities of the idea of local control.

Consider, for example, the lessons of the famous Cornell Peru Vicos Project, which had local self-sufficiency as its goal. In the 1950s anthropologists from Cornell University and the Peruvian National Indian Institute leased a hacienda for a five-year period and proceeded to initiate changes that empowered the peasants through the development of democratic decision-making processes, educational and health facilities, and improved cultivation and marketing procedures. There is now a huge literature on Vicos, including many debates, and the project was far from uniformly successful. For example, it tended to cause increased inequality between men and women, as is often the case with development projects that allow men greater access to the cash economy. Notwithstanding failures of this sort, the demise of the project was related to macropolitical events rather than internal shortcomings. As news of the project's success spread, other peasants began to call for similar innovations. Landowners in the valley and throughout the country soon intervened to stop what they perceived as a subversive project that threatened the existing class structure of the country. The Vicos project was therefore revealed as a model of land reform, and calls for land reform in many third world countries are equivalent to calls for armed revolution.

The Vicos Project focused on peasants in the national society, so it inevitably trampled on the toes of large landowners and precipitated a backlash. The rise and fall of Vicos indicates how important the control of the land is to any form of grassroots development project. Although Native peoples usually have control over some land (often much less than they originally had), territorial integrity is a first step to grassroots development. In some cases indigenous peoples or peasant farming communities must work with national and international organizations to gain the leverage needed to pressure national governments into demarcating and protecting their land.

However, the coalition politics entailed in struggles for land rights may also mean redefining land rights struggles as environmental struggles, which are more marketable among the progressive segments of the urban middle classes. That strategy in turn entails new risks. For example, in some cases the resulting debt-for-nature swaps or biosphere reserves can actually mean a loss of local control over land use. Peasants in one such area of Central America have faced such stringent regulations that they cannot even practice their traditional slash-and-burn agriculture or fell trees without permission from the regulating agency. Here international environmental groups show such insensitivity to longstanding local needs and practices that they end up playing the role of new imperialists.

Another complicating factor in the development of local control is the problem of divisions within the Native society. Like other governments, tribal councils may grow distant from the people and discourage community participation and democratic action. The process of estrangement and polarization within the Native society may occur in the context of ostensible empowerment through control over revenues generated by mineral and other extractive concessions. As the well-known case of the Navaho and Hopi demonstrates—and apparently a similar process is beginning to occur among the Kayapó—energy development projects may bring about alliances between outside interests and indigenous leaders, thus exacerbating inequalities within the indigenous population and creating tensions with neighboring populations as well. The Vicos case is also typical of the general tendency for development projects, even well-intentioned grassroots projects, to exacerbate gender inequalities. Gender inequality is not a necessary outcome, but the projects need to protect against this eventuality by setting up credit unions and access to the cash economy for women.

The anthropologist Allen Turner's work (1987) among the Kaibub Paiute of northern Arizona provides one model of how an anthropologist (or other concerned outsider) can help reinvigorate direct community participation and democratic processes within the tribe. With the support of the tribal council, Turner held weekly meetings that solicited direct community input toward the construction of a tribal plan. The meetings eventually helped mobilize the tribe to such an extent that Turner described the process as a social transformation that involved not only more egalitarian relations within the tribe but also a much more assertive and successful stance with respect to the federal government (1987:126–27). At the recommendation of the planning program, the tribal council created a department of plant management, which in turn initiated and supervised several technical improvements, such as upgrading the water supply system, repairing irrigation systems, installing a pipeline and storage tank system, and setting up a sanitary landfill.

Although some of the energy for direct citizen participation apparently dissipated after the project ended, Turner concludes that "the Kaibab Paiute Tribal Council is today more receptive to community ideas" and that it is also "much less reluctant to accept any or all federally sponsored programs without assessing their benefits and costs to the community" (129).

Turner's project was of relatively short duration, but in other cases anthropologists have developed long-term working relationships with local communities. A good example is the work of the anthropologist John H. Peterson with the Mississippi Choctaws, which he describes as follows:

> During the past sixteen years, my role as an anthropologist changed from that of a neophyte fieldworker to a middle-aged administrator at a nearby university. My positions with the client group included working as an externally funded fieldworker, unpaid friend, paid consultant, full-time tribal employee, subcontractor on tribal programs, and appointed member of the Choctaw Heritage Council. The client group changed from an Indian tribe just beginning to question the complete authority of the almost totally white-staffed Choctaw Agency, a branch of the Bureau of Indian Affairs (BIA), to an aggressive tribal government with major tribally owned industrial enterprises and an Indian-directed Choctaw Agency. Under the circumstances, my role as an anthropologist shifted in accordance with my own knowledge and capabilities, the needs of the Choctaws with whom I was working, and their knowledge and capabilities for utilizing my anthropological skills and talents. (1987:264)

Peterson was not responsible for the changes he describes; instead, he provided technical and other forms of assistance to the Choctaw in ways largely defined by the tribal council. His work brings out the importance of developing a flexible, long-term relationship: rather than appear on the scene for the brief period of a development project, the anthropologist or activist becomes part of the community and changes roles over time in response to the needs and decisions of the community.

The examples of Vicos, the Paiute, and the Choctaw help circumscribe some of the parameters needed for defining and achieving local control. Vicos shows the importance of support, or at least noninterference, from the national government. For Native peoples governmental support involves de jure and de facto recognition of land rights. (In the United States there has been a longstanding policy of just the opposite: termination of Native American tribes and reservations.) Thus, like resistance struggles, grassroots development is very dependent on the position of the national government. Most of the model projects are located in democratic countries or in large national societies where there is some commitment to helping indigenous peoples. It would be foolhardly to talk about locating grassroots development projects in countries where the national governments are engaged in genocidal policies, either through direct intervention or through malignant neglect.

The Paiute case raises the difficult issue of equity and participation within the tribe. The issue is difficult because there is a trade-off between imposing Western forms of democracy on a non-Western society and helping oppressed categories within the non-Western society, particularly women, achieve greater control over their lives. (I have outlined my own approach to the dilemma in the discussion of the Yir Yiront above, but dilemmas of this sort need to be approached on a case-by-case basis.) Finally, the Choctaw case provides a longitudinal perspective on the question of local control, and it shows how flexibility over time is an important part of grassroots development. Peterson shows how anthropologists—as symbols of model citizens of the national society—also undergo their own form of grassroots development alongside the developments of the people with whom they are working. Anthropologists such as Peterson also contribute to redefining their profession in ways that breaks with a past that was often heavily compromised by colonialism.

Technology and Grassroots Development

Having now outlined some of the difficulties and parameters of the idea of local control in grassroots development, I will end by focusing on the question of technology and political economy from a comparative perspective. My discussion provides one way of thinking about the issue by cross-secting type of technology and relationship to the global economy (see table 8.2).

The simplest form of grassroots development project involves low (or

TABLE 8.2
A Typology of Technology and Grassroots Development Projects

	LOCAL USE	NONLOCAL USE
LOW OR LOCAL TECHNOLOGY	Ladakh project, Gandhian-style engineering projects	Huichol carpentry, Brazil nut products, alternative paper production
HIGH OR NONLOCAL TECHNOLOGY	Haitian reseeding project, Shuar radio project, Western medical projects	Pharmaceutical production, indigenous media

local) technology for local use. One example is the Trombe wall used in the Ladakh Project in the Little Tibet region of Nepal. Made from a double layer of glass attached to a wall made of local stone and mud brick, the Trombe wall takes advantage of the sunny climate of the region to provide a system of passive solar heating. Because local resources are used, the wall does not create relations of dependency on the outside world, and it cuts dependency on fossil fuels such as coal. The linguist and development activist Helena Norberg-Hodge calls the project an example of counterdevelopment, in which local initiatives are used for sustainable alternatives.

In India there are many projects of a similar type that are inspired by Gandhians. Although Gandhians have been criticized from the left for being politically naive and unable to organize successfully against large-scale macrodevelopment projects, they have been very successful at designing a number of grassroots development projects. For example, one grassroots project in India provided a locally controlled solution to the water problem. The increased use of electrical pumps has lowered the water table, and high-tech reservoirs bring with them many problems, among them loss of land and local control. Instead, the villagers solved their water problem by building a series of small ponds that provide local access to water during droughts. Another example of a Gandhian-inspired project that uses local technology for local use is one of the many projects of the abovementioned Shri AMM Murugappa Chettiar Research Centre (MCRC) in Madras. Researchers at the center such as the engineer C. V. Seshardri are guided by Gandhi's teachings, which emphasized nonviolent lifestyles anchored in rural, agricultural communities. One of their projects provides a low-tech solution to rural Indians' problem of vitamin A deficiency, which leads to numerous cases of vision problems and even a condition that causes blindness. The standard solution to the problem has been to turn to international pharmaceutical companies and to import the requisite vitamins. However, for a number of reasons that solution is not satisfactory: drug supplies do not always make it to the people, the companies often use animal testing (which Gandhians and others find inhumane), and the expensive import costs for their products only make the nation more dependent on international companies and rural communities more dependent on the Indian government. The MCRC worked out a simple and effective way for families to cultivate algae in vats in their back yards. A dietary supplement of spirulina algae provided enough vitamin A to overcome the deficiency. The vats can be made from local materials or they can be made out of cement provided by the government. In either case, the system is a sustainable source of vitamin A that does not harm the local ecology or require dependency relations on the outside world.

A second type of grassroots development project involves outside, high technology for local use. Many medical projects fit in this category, especially those that train local people to take over medical care. Likewise, Native use of infrastructure technologies fit into this category. An example is a development project for the Shuar of Ecuador. A $40,000 grant from Cultural Survival and the Inter-American Foundation allowed the Shuar of Ecuador to avoid becoming caught in the Spanish-language national education system by using radios to develop education in the Shuar language from a Shuar perspective. In many cases indigenous peoples are using video to record rituals and events and thus to preserve their culture, and in some cases they have their own time on local radio or television.

A well-known example of nonlocal technology for local use is the Haitian reforestation project headed by the anthropologist Gerald Murray ([1987] 1989). Drawing on his knowledge of the ecology, national politics, and peasant economics of Haiti, Murray designed a successful reforestation project that allowed peasants to make their own decisions about the planting and harvesting of trees. I classify the project as nonlocal, high technology rather than local and low because the seedlings were of a special type that had to be prepared in nurseries; however, once planted the trees could become local, sustainable resources. Whereas previous, top-down reforestation projects had been dismal failures, this project turned out to be more successful than even Murray had originally envisioned. Part of the success involved technical factors. For example, Murray chose fast-growing wood trees that could be harvested within four years for charcoal and construction needs. The trees restored nutrients to the soil, and a nursery developed a much smaller microseedling that could replace the bulkier bag seedlings used previously. The planting strategy also allowed for the "physical juxtaposition of trees and crops" rather than consigning an area of land only to trees, and this arrangement provided greater crop security for the peasants. However, while these technical factors were important elements for the success of the project, the key was that control over planting and harvesting decisions was ceded to the peasants. Once they discovered the benefits in the project, many peasants soon joined in.

A third type of grassroots development project involves developing local technology and products for the national or global economy. One of the projects of the MCRC has been to develop a sustainable alternative to wood pulp as a source of paper. Paper products come primarily from softwood trees, which are plentiful in the northern countries of the northern hemisphere (Canada, Sweden, etc.) but rare in India. As a result, the country has a severe paper shortage. To make the paper, the MCRC has experimented with the production of silk-cotton trees, which produce a pod that can be harvested without killing the plant. The pod is made of cellulose, and it can be processed to make paper. (The MCRC has also developed other alternatives to paper; for example, the business card that C. V. Seshardri gave me was made of a plastic that was derived from sugarcane alcohol.)

Several grassroots development projects fall into this category. For example, the Irulu of India developed a snake farm in which they milk poisonous snakes and sell the venom to laboratories that make an antidote serum. The Huichol of Mexico, with the help of an anthropologist and a forestry expert, took over a timber concession that was cutting down the tribe's best trees, and they started their own carpentry industry. Writes Clay, "[T]hey have even encouraged reforestation by selectively cutting damaged and diseased trees" (1989:55). Perhaps the best-known of these projects is developing a Brazil nut cooperative in the Amazon that markets the nuts to Western businesses, including Ben and Jerry's Ice Cream, which now makes the flavor Rainforest Crunch. The Body Shop is another example of a first world firm that is helping to create a demand for rain forest products, and the company has worked with Cultural Survival to help educate consumers about not only the rain forests of the world but also the indigenous peoples to whom many of the forests belong.

By creating linkages between indigenous enterprises and national and global markets, activists hope to create economic incentives for preserving rain forests and other habitats of Native peoples. If national governments can be convinced that, for example, a forest is economically more valuable than a ranch, then pressures on forest dwellers may be mitigated. At the same time the new export-oriented industries allow Native peoples to begin to enjoy some self-sufficiency through economic self-development consistent with their traditional way of life, rather than being pulled into factories, mines, and ranches as wage laborers.

However, there are dangers to this approach. Consider the following possibility. Activist programs are so successful that global consumer consciousness creates a huge demand for rain forest products. Many companies, not just progressive companies such as the Body Shop or Ben and Jerry's, decide to go rain forest green. MacDonald's makes Brazil nut hamburger buns, Kellogg's makes Brazil nut granola, and Skippy makes Brazil nut butter. In the quest for profits, indigenous cooperatives and enterprises are sidestepped for debt peonage systems, and we are back to the days of rubber tappers. The Amazon undergoes another of its boom cycles, this time with Brazil nut barons rather than rubber barons. Then companies find a way to bioengineer Brazil nuts or to grow them in plantations in Asia, and they mix genuine Brazil nuts from the Amazon with plantation-grown Asian varieties and bioengineered, laboratory-grown cells. Just as the Amazonian economy crashed before, when the British smuggled out the rubber trees and set up plantations in Asia, so the Brazil nut business goes bust.

The scenario may sound ludicrous, but it is not impossible. For that reason, I think the scenarios of grassroots development need to consider other options. The fourth type of grassroots development is high technology that provides income through export. The two main examples at present are locally produced media for export and pharmaceutical research that harvests tropical plants for tests for potential use as drugs. At present the radio and television stations of Australian Aborigines are one of the few cases of media production controlled by indigenous people but consumed by a broader population. In central Australia the Aborigines run one of the most widely listened to radio stations in the Northern Territory, and they also control a television station that broadcasts in six aboriginal languages and in English. In North America Native American-produced media is now available through a few rental agencies, although generally for educational purposes at this time.

Indigenous art and music is a related example of a mediated, high-technology form of production that is exportable for nonlocal consumption. The birth of world music has brought indigenous music to prominence in music stores and on the air waves. Likewise, some indigenous peoples have transferred their artistic traditions onto the media of contemporary art such as acrylic paint, and their work has entered into circulation on the international art scene.

The production of indigenous film, video, music, and art for nonlocal consumption is likely to involve complex negotiations both between indigenous and cosmopolitan worlds and within indigenous societies. The anthropologist Terence Turner (1992) notes that for the Kayapó the role of cameraperson can lead to a powerful position as culture broker, and it may be one path for ambitious young people to work their way to leadership positions. Likewise, some scholars have criticized indigenous media and art as selling out local traditions or risking appropriation by the world system. Those criticisms, as Turner points out, should be taken with a grain of salt, for they may lead to conservative measures that limit the rights of indigenous people to have access to the cosmopolitan media technologies and markets.

As discussed in the previous chapter, the production of pharmaceutical products from local plants will only work if accompanied by honored royalty arrangements and protections on indigenous people's intellectual property rights. Setting up guidelines and developing international regulations on intellectual property rights are first steps. After those steps are taken, indigenous peoples should also be able to receive training in the science and technology that would allow them ultimately to be able to produce their own medicines and cosmetics.

In conclusion, successful grassroots development in the global economy will have to entail local control of high technology—and not merely its use, but its production. Native peoples must be seen not only as the suppliers of ethnopharmacological lore or sites for exotic films and music but ultimately as those who control the means of their production. I would even go so far as to argue that a long-term goal of grassroots development should be Native-controlled universities that could train indigenous peoples to enable them to control production in high-technology fields, should they desire that option. It should be clear that I am not speaking of imposing institutions on Native peoples who do not want them. Rather, I am advocating support for Native efforts to develop their higher educational systems and cooperative enterprises. The goal may sound utopian, but there are already Native colleges in the United States, and it would not be a huge step to develop Native-controlled media and pharmaceutical cooperatives from the Native colleges. Only through control over higher education and high technology can indigenous peoples and other rural communities have a chance at access to the intellectual and financial resources they need to control, protect, and preserve their lands and cultures and to resist the ethnocidal and ecocidal forces that surround them.

9 | Conclusions: Science, Technology, and the Multicultural Education

The Hudson Institute *Workforce 2000* report includes one of the most frequently cited—and miscited—statistics of the 1990s: by the year 2000 only fifteen percent of the new additions to the American labor force will be native-born white males. Forty-two percent of the new entrants will be native-born white females, and the remainder will be immigrant and nonwhite men and women. I use the statistics for a way in to a conclusion that speaks to the issue of multiculturalism, science, and technology in the context of higher education.

The Hudson Institute report has now taken on a life of its own as it has circulated through the media, the corporate world, and the universities. The fifteen percent statistic seems to be the 1990s version of the 1960s Bob Dylan song about Mr. Jones, who knows that something is happening, but is not sure exactly what. The nature of the report as social science has been dwarfed by the general cultural phenomena that the statistic symbolizes: the emerging demographics predict that by the middle of the twenty-first century most Americans will trace at least some of their ancestors to a continent other than Europe. In the United States, as in many other Western countries, native-born white males today realize that they are going to have to work with women, nonwhites, and immigrants; they are even going to work *for* them, if they are not already doing so. For decades some, but clearly not enough, white males have fought alongside women and members of underrepresented groups—often courageously and often at a personal loss—for sharing power on moral grounds. In the closing years of the twentieth century, the moral question of what *should* be happening is shifting to a factual question of what *is* happening.

The famous fifteen percent statistic has sometimes been miscited to the effect that only fifteen percent of the workforce in the year 2000 will be white males. The miscited version of the statistic seems to have provoked outbreaks of hysteria among some white males. Even the properly cited version can be unsettling enough to provoke a circle-up-the-wagons kind of response. For example, I was once talking with a group of potential university donors—all older, white, male industrialists—when I mentioned the statistic. One denied that it could be possible. Another replied, "What are we going to do about it?" Another was quick to answer: "Well, our company is a United Nations of people from all over the world: men, women, foreigners, blacks, whites, Hispanics. We all get along because we all treat each other as individuals."

Beyond Recruitment and Retention

The response of the third industrialist is problematic in a number of ways. First, it is likely that the daily social relations of an organization are full of incidents of sexism, racism, and so on that have gone underreported and unrecognized. Those incidents may not be visible to white, male leaders unless they start to listen. Second, the ideology of a "United Nations" utopia may actually prevent bias incidents from coming to light, because reporting such incidents would mean questioning an official organizational culture. Finally, the idea of "treating each other as individuals" begs an important question: Individuals on whose terms, or according to whose definitions? It is easy to let in other groups if, for example, cultural differences in communication styles are denied under a rhetoric of individualist equality defined by dominant groups.

The continued existence of glass ceilings suggests that society is a long way from achieving the individualist equality that some people would like to claim exists in their organizations. Women and underrepresented ethnic groups have made strides in some of the high-status professions, including science and engineering, but the statistics show that gender and ethnic gaps are still huge in those fields. In engineering, for example, the ratios are as high as one woman to every twenty-five men. Something may be happening, but it ain't happening very fast.

Discussions of multiculturalism and cultural approaches to science and technology are therefore part of an ongoing historical process of modernization in the academic and professional world. One of the ways in which ascribed hierarchy continues to operate in relatively modern and progressive societies such as the United States is the continued exclusion of women and people of color from the technical professions. Although Asian Americans have been very successful in the technical fields, recruitment and retention for other historically underrepresented groups and for women in general has been poor. Among faculty and students at the nation's top engineering and science schools, men outnumber women by ratios such as four-to-one or seven-to-one, and percentages of African American and Latina/o students and faculty lag far behind their already low numbers in other disciplines. Likewise, studies have shown that even in the 1980s, when the wage gap between American men and women narrowed after decades at the infamous figure of fifty-nine cents, women in engineering have continued to experience significantly lower promotions and salaries.

Thus, at the first level the issue of multiculturalism, science, and technology involves recruitment and retention: how can more women and members of underrepresented ethnic groups be encouraged to join the technical professions? That issue is now the topic of a great deal of thought and policy action among faculty and university administrators in the United States and Canada as well as in other countries. The many people who are working on the issue realize that recruitment needs to extend into the elementary schools, where the sometimes subtle effects of cultural stereotypes already begin to sort out boys and students of dominant ethnic groups as more successful in math and science. Furthermore, recruitment needs to extend to faculty to provide diverse role models for the students. The many reformers also realize that recruitment efforts need to be balanced by retention programs that keep students in technical college-level programs once they have been admitted. Those programs can range from tutoring and mentorship to overhauling the "boot camp" mentality of some engineering schools. Some schools are looking at ways to reform course sizes, test procedures, grading on the curve, and any number of other features of the masculinist boot camp culture of the science and technology curriculum. In other words, they are beginning to link recruitment and retention to the overall culture of science and technology disciplines. As the anthropologist Gary Downey and colleagues have shown in a detailed study of the weeding out process in engineering, the engineering student is socialized into a regimented and disciplined lifestyle that may be especially foreign to women and members of some underrepresented ethnic groups.

All efforts to increase equality and diversity through recruitment and retention of students in the technical fields are very important in the struggle to break through the glass ceilings that hold back certain groups of people. My concluding comments extend and compliment these efforts by focusing on the related question of curriculum reform. As has been the focus of the entire book (and the new social studies of science and technology), my discussion is meant to ensure that the debates on multiculturalism and science/technology education do not forget the question of content. To many scientists and engineers the question is anathema if not nonsensical or heretical because the content of science and technology—theories, methods, and technical designs— is assumed to be socially neutral and above or beyond culture. However, this book as well as the research upon which I have drawn point to numerous ways in which at the very minimum that position is untenable in many circumstances.

Thus, discussions and policies on multiculturalism and the science and technology education should not be limited to recruitment and retention, however important that basic issue is. By limiting the discussion to that question, it is possible to fall into the trap of the third industrialist mentioned above, who believes it is possible to admit women and members of underrepresented groups under a mantle of egalitarian individualism in which nothing else changes. Instead, the many examples that I have discussed in this book point to a different sort of situation: when the social body is diversified, the content of science and technology changes as well as its communication and social structures. When an African American entered agricultural research, he came up with new theories or criticisms of existing ones; and when a woman was allowed to design a car, she came up with new features that men had never considered. In general, when underrepresented groups gain power and voice, institutional policies and interactional practices are likely to change. Thus, it is not simply a question of diversifying the technical professions at the level of who gets to practice them; it is also a question of recognizing that when a diversification of bodies occurs so does a diversification of knowledges, designs, and practices.

Steps Toward a Multicultural Science and Technology Education

University curriculum and classroom practices are only two small pressure points in the struggle to build a university and a society that celebrates and honors difference as a source of vitality and creativity. However, postsecondary science and technology education is an important pressure point because it is a gatekeeper to many highly paid professions. As I have shown in this book, science and technology are also sites for the reproduction of an ideology that values some groups and cultures over others. Thus, alongside the issue of recruitment and retention we need to think about multiculturalism in the science and technology curriculum. Solutions must emerge locally and be crafted to local needs, resources, and populations. Although I will not offer a single, universal solution, I will suggest some possibilities.

An initial strategy is to begin to think about integrating a social science or humanities requirement into the science and engineering major in a manner that goes beyond standard distribution requirements to take a certain number of humanities and social science courses. For example, majors in technical fields could be required to take a course on the historical, social, cultural, or ethical aspects of biology, engineering, or one of the sciences. If the university has an STS program, then the technical curriculum might require that the humanities and social science requirement include an introductory STS course and/or another STS course oriented toward ethics, values, and diversity issues in science and technology. Another good place for cooperation and bridging between the STS and science/engineering curricula is in jointly sponsored lectures and colloquia. Guest speakers also provide an opportunity to bring in role models for women and underrepresented ethnic groups.

This first step may require more changes from the STS side than from the science and engineering curriculum, because many STS programs focus on historical and philosophical approaches that tend to exclude questions of culture, politics, and values. Yet, it is also possible to use existing resources and build an STS program with a diversity and ethics-and-values orientation. To give one example of how the humanities and social sciences curriculum can be reformed, at Rensselaer I have codirected a project sponsored by the Fund for the Improvement of Post-Secondary Education to build a humanities and social science minor that broadens the STS curriculum to include cross-cultural issues. I teach an introductory course that explores many of the topics raised in this book, and a number of faculty teach area studies courses on science and technology in their global region of speciality. We were able to put the program in place by using existing courses and only limited new course development. Faculty with area studies interests now offer courses on the Arab world, India, Japan, Latin America, and so on, and they include science, technology, and medicine components in those courses. We have also developed a capstone course in which students integrate a cross-cultural/multicultural perspective into an engineering design project or participate in a field project with an ethnic group different from their own. Although still fairly new, the program has achieved substantial student interest. The faculty on the program were drawn mainly from STS, but interested humanities and social science faculty from other departments were also involved. Our cross-cultural studies minor is only one example of any number of minors that could be developed, depending on local resources. Many combinations are possible: "women's studies and biology," "comparative environmental studies," and so on. One useful resource for this type of program is the book *Biology and Feminism*, in which the biologist and feminist science studies researcher Sue Rosser has articulated six phases in a program for curriculum change in biology and women's studies.

The first-level solution of developing linkages between the STS and technical curricula is likely to meet with relatively little resistance from the science and engineering faculty because their curriculum is complemented rather than transformed. The only struggles that are likely to emerge from that solution will be struggles over resources, if they are perceived as existing in a zero-sum relationship and if changes occur through expansion rather than reorganization. However, the first-level solution represents only one way in which multicultural issues and perspectives can be brought to students in the technical fields. An additional step is to integrate introductory science and engineering courses with introductory STS courses on the same topic through interdisciplinary, team-taught courses. In other words, students learn technical skills in interdisciplinary courses that simultaneously locate those skills in their culturally specific sites of production. For example, introductory college-level courses in mathematics could be taught to include the cross-cultural history of mathematics and ethnomathematics. Instead of taking two one-semester courses (one in math and one in STS), students would take a two-semester sequence on math/STS in which they learn the same content for the same credits but in a more integrated fashion. Environmental studies is another area where integrated interdisciplinary approaches are likely to be successful. A less-ambitious alternative is to set up special sections related to diversity issues (such as sections on women in science) in large introductory courses, a practice that has been instituted in some universities.

Even where there is no STS program in place and there are no resources for interdisciplinary team teaching, it is possible for scientists and engineers to teach their own courses from an interdisciplinary perspective. That is easier in applied courses—such as courses on environmental studies—but it is possible in even the basic science curriculum. For example, concepts and theories that bear biographical names can be introduced by providing brief histories of the people behind the names. It makes lectures more interesting, does not require a great deal of extra preparation time, and can point to diversity issues in science and technology. The change seems trivial, but students say it is important as a source of inspiration, especially when the exemplars include women and people from outside North America and Northwestern Europe. As one African American student said to me, "Even if it's an Italian or Russian, at least I'm seeing they're not all American and English white men."

Yet another direction is to provide funds for faculty development so that the science and engineering faculty themselves can develop introductory courses to include comparisons with non-Western systems of knowledge. In chapter 7 I have provided a sample of and points of entry into the literature for most of the ethnoknowledges. For almost any introductory science or engineering course, it would be possible to begin by examining non-Western constructions of the field of inquiry or the technology. Throughout the course it might be possible to introduce non-Western points of comparison. In addition to making a technical course more interesting, the practice might also help clarify some of the basic assumptions of the Western knowledge that is being taught. Non-Western perspectives can also be a useful starting point for the interdisciplinary analysis of grave global problems such as global warming, pollution, starvation, drought, and overpopulation.

Finally, multicultural issues can be worked into the curriculum through the use of examples. The "story problems" of problem sets are more than just techniques for teaching students the technical content of the course material in an applied setting. The examples and concrete details often come from textbooks or professorial experience, and these sources are frequently still quite Western, male, and/or white. It can be interesting, and instructive, to develop story problems set in other countries or against the backdrop of ethnic and gender differences. It can also be helpful to point out and discuss the choices of metaphors in the standard textbooks that are being used, providing students with critical reading skills at the same time that they are learning basic science or engineering. For example, in *The Woman in the Body* and some of her other essays the anthropologist Emily Martin provides some useful examples of ways to read metaphors used by scientists in diagrams and descriptions of biological processes.

Several scholars, perhaps most notably the anthropologist Sharon Traweek, have also advocated teaching "corridor talk" as part of the classroom education. The knowledge about what it means to become a member of a technical profession often gets passed along outside the classroom in individual interactions among faculty and students. If those interactions tend to favor male/male or white/white relationships, then some of the crucial know-how of becoming a scientist (or engineer, etc.) remains out of reach for the unfavored students. Traweek and others have advocated moving into the classroom the informal education of the science and technology profession: knowledge about passing through career stages, building a CV and networks, who belongs to whose network, how to find grants, and so on. Those issues could also be taught via interdisciplinary courses in which the history and sociology of the profession are taught together with the field's technical content. Although astute scientists can teach their own history and sociology well, they might also benefit by team teaching with members of the STS faculty who specialize in their discipline.

Other issues related to the science and technology curriculum involve classroom practices. Faculty training workshops can help sensitize faculty to intercultural communication differences and a variety of other ways in which unconscious bias enters into their own and students' classroom behavior. Many of the communication issues stem from patterns of socialization that students carry with them into the classroom and that professors can learn to recognize and change. Those socialization patterns interact with the Pygmalion effects of the faculty, such as the documented tendency to ignore women students and students of color and instead to call on white or male students. Those communication patterns run very deeply and are not easy to change. For example, I found that even after using one of the books of the linguist Deborah Tannen as a text and calling attention to the male "airspace" problem, the men continued to grab airspace. To mitigate the problem I have experimented with a number of techniques, such as pointing out the classroom dynamics when they begin to operate, raising the issue explicitly and asking the students what they think should be done (if anything), not calling on the first man and sometimes waiting until a woman raises her hand, or dividing up the classroom into small groups, sometimes by gender. It is important to recognize that we are all learning; I certainly am, and I continue to experiment with new teaching practices each semester.

A number of techniques in interactive learning can also help make the classroom a more hospitable environment for students of all backgrounds. Small-group discussions in which the teacher remains at some distance can provide a more relaxed environment, especially when the groups are given a cooperative task. In many cases laboratory projects are set up on a competitive basis, and they can be changed to cooperative group projects. At the same time, laboratories can be a location for unwelcome sexual advances and other forms of discriminatory behavior, so they warrant careful attention. Engineering design projects present another opportunity for building diversity through cooperative learning. A survey of students at the University of California at Berkeley revealed that they did not think cultural sensitivity training sessions or interethnic parties were particularly effective ways of promoting campus diversity. The students found more fulfilling and less contrived those class exercises in which classmates from different ethnic groups worked together on a joint project. A related idea is to set up senior projects, such as the capstone engineering project or senior thesis, so that the advisor includes a social scientist or humanist and the project integrates a social or cultural perspective into its research design. Finally, in some universities, such as the University of Cincinnati program set up by the engineer Kenneth Challenger, co-op programs have become internationalized, so that engineering students have an opportunity to work abroad and experience cultural differences through immersion.

By diversifying the curriculum as well as the student body, colleges and universities can teach the underlying message that each generation reconstructs its own science and technology. As the new generations diversify, so will their science and technology. Cultural analysis of science should not remain the province of social scientists and humanists;

rather, it should become part of the heritage of members of all technical professions. My hope in writing this book is to provide a number of ways for scientists, engineers, and technical educators in general to think more culturally about their own work. Too frequently we have been taught to see culture as a corrupting force in the production of science and technology, and to see anthropological, sociological, and historical approaches as airing the "dirty laundry" of how social forces have impinged on a field of human endeavor that should be somehow asocial. I hope instead to have contributed to an alternative perspective that sees cultural politics not as extraneous to the content of science and technology but instead as a framework for a different way of seeing the same content. A cultural-and-power approach provides one framework for allowing in new and different voices and, with them, ultimately new and more just research and technologies, and a better society. It is opportunity, not a threat, that is knocking at the door.

Notes

1.Introduction

The field of science and technology studies, although interdisciplinary, owes a great deal to what is sometimes called the new sociology of science or social studies of knowledge. In contrast to the old sociology of science characterized by the American sociologist Robert Merton, as well as the old history and philosophy of science, the new sociology of science took as its object of study the content of science and technology, that is, questions of theory construction, technical controversies, and design issues. Key texts and viewpoints in this movement are the articulation of the strong program (Bloor 1976), the historical studies of the Edinburgh school (e.g., Barnes and Shapin 1979), the laboratory ethnographies (e.g., Knorr-Cetina 1981, Latour and Woolgar 1979), the Bath school of relativism (e.g., Collins and Pinch 1982), reflexivity (e.g., Woolgar 1988b), discourse analysis (Mulkay, Potter, and Yearley 1983), social epistemology (Fuller 1993), and the actor-network theories (e.g., Latour 1987, Callon, Law, and Rip 1986). Not all of those theoretical positions can be subsumed under the term *social constructivism*; often the term is associated with the laboratory ethnographies. However, some version of social shaping or social construction is by definition central to the idea of a social studies of knowledge.

Because I mention actor-network theories more than once, it may be worth explaining that this influential framework explains the success of scientific "discoveries" or technological innovations in terms of network building. The networks are heterogeneous; that is, they include people, machines, texts, and institutions. The validity of a fact or success of an innovation is attributed to the ability to build and maintain a strong network and to "enrol" *[sic]* others in one's network.

The main theoretical tendencies of STS, as well as debates among them, are represented in Knorr-Cetina and Mulkay (1983) and Pickering (1992). Laudan (1990) provides an overview of the different philosophical positions of the relativist, realist, positivist, and pragmatist; however, his account is clearly set up to privilege his own pragmatist position, and it is far from the last word on the topic. A somewhat more advanced survey of positions in contemporary philosophy of science as informed by STS is found in the interviews by Callebaut (1993). What I have called the new sociology of science is a field of debate and cocitations that tends not to include the perspectives and arguments of feminists, radical political economists, cultural studies scholars, and cultural anthropologists. For an early version of an attempt to articulate an alternative to the new sociology of science, primarily at the level of methodology, see Hess (1992b), and for a more complete discussion of anthropology in STS, see Hess (n.d.). For general overviews of STS that include feminist and cultural studies perspectives, see Traweek (1993) and Rouse (1991), and Pfaffenberger (1992) for a review of the anthropology of technology. For the examples on physics mentioned in the introduction, see Traweek (1988) and Hess (1994a).

Although I have signaled the influence of Foucault in my definition of power, I do not claim to be working with a Foucauldian notion of power. Fraser (1991) argues that three distinguishing features of Foucault's concept of power are that it is productive rather than prohibitive, capillary rather than statist or economistic, and practice-oriented rather than belief-oriented. In another post-Foucauldian discussion of power, Law (1991) distinguishes power to, power over, power/storage, power discretion (the power not to do something), and power effects. I am concerned most with the question of effects, and inasmuch as one wishes to conceptualize scientific theories as beliefs rather than discursive practices and technologies as embodied theory rather than rationalized practices, I am concerned with both belief/discourse and practice. A related and influential perspective in anthropology has been the project of "anthropology as cultural critique" (Marcus and Fischer 1986) associated with the Rice school, which includes several students and faculty now making contributions to cultural studies/feminist/postcolonial anthropologies of science and technology.

On the question of racial/ethnic terminology and categories, there are several fine essays in Harding (1993), among them those by Marshall and Livingstone. The same argument could be applied to the classification system for gender and sexuality, which is undergoing radical reconceptualizations as well (see Rubin 1993). The interaction of the two system produces further complexities; for example, is a WASP gay man a minority?

2. The Cultural Construction of Science and Technology

My discussion of technototemism follows the long sociological tradition stemming from Durkheim ([1915] 1965) and Durkheim and Mauss (1963). In the STS literature this tradition has been investigated (e.g., Bloor 1982, Shapin 1979) as well as criticized (Latour 1990). A major criticism of conventional uses of Durkheimian/Maussian social theory is that it takes social categories for granted and sees natural categories as a projection of social categories onto a natural world, with social categories as the cause and natural categories as the effect. Other versions take natural or social categories for granted and see the problem as one of mapping the two onto each other. Latour argues instead for a perspective that sees both social and natural categories as coconstituted. I favor this view as well and would argue that a similar viewpoint is found in the interpretation of totemism put forward by Lévi-Strauss (1963b). Just as the linguist Ferdinand de Saussure argued that sound and meaning were coconstituted in the sign (or trace/signifier in poststructuralist readings), so society and nature are coconstituted in the totem.

In formal terms, Y1 is to Y2 as X1 is to X2, where Y1 and Y2 are distinctions in the natural code and X1 and X2 are distinctions in the social code. Clearly, multiple categories and codes are operative in case studies. Although Lévi-Strauss is perhaps most closely associated with formulas of this type, the general method of interpreting intellectual divisions by showing how they map onto social divisions is not limited to any one school within cultural anthropology, sociology, literary-cultural studies, or history. Nor should the method be seen as restricted to modernist social theory. Although I have used a modernist way of presenting cultural constructivism (through a closed formula), one does not need to think of totemic relationships in terms of discrete, formal systems. One need only study, and read with care, the careful keeping tabs of the totemic relations of class, gender, ethnicity, and national culture in the work of a "postmodern" culture critic such as Haraway to understand that the totemic principle is transgressive of the modernist/postmodern divide. Likewise, see Bourdieu's *Distinction* (1984) for a development of the analysis of totemic relationships in the context of capitalist societies.

On the Darwin/Marx correspondence, see also Hyman (1974:122), and for a related discussion of Darwinism and Social Darwinism, see Shapin and Barnes (1979). A frequent but erroneous view is that Marx wanted to dedicate *Das Kapital* to Darwin (Carroll and Feuer 1976). The letter I have cited provides a better sense of Marx's opinion.

On the example of statistics in Britain, MacKenzie makes a more convincing argument for the connections with class in a chapter comparing the biometricians and Mendelians (MacKenzie 1983, chap. 6). On the imputation debate, see Woolgar (1981a, 1981b), Barnes (1981), Barnes and MacKenzie (1979), MacKenzie (1978, 1981, 1983:chap. 7); see also MacKenzie (1984) and Yearley (1982). Although my example of technototemism and class draws on statistical measures, numerous other examples are possible. One example is Cowan's "How the Refrigerator Got its Hum," which shows how the electric compression refrigerator defeated the gas absorption refrigerator (1985a; see also 1983). Although the latter had numerous technical advantages, it was not aligned with the large corporations and the emerging electric and electric motor industries. For the formulation of social shaping/social effects as a "seamless web," see Bijker and Law (1992).

On Ernest Everett Just, see Gilbert (1988, 1991) and, for his biography, Manning (1983). Manning (1983:263) argues that Just used the cell as a model of cooperation in which one could seek the "roots of man's ethical behavior." Gilbert (1988) also contrasts Just to his contemporary, Richard Goldschmidt. Just eventually moved to Europe because of the racial discrimination that he experienced in American universities, just as Goldschmidt, a German Jew, moved to the United States after the Nazi rise to power. However, Goldschmidt's cellular map emphasized the importance of the nucleus, and Gilbert argues that Goldschmidt's views correspond to his position as a member of the intellectual German Jewish aristocracy (see chapter 5 for my discussion of mandarins and pragmatists based on Harwood's research). Another direction of contrast involves the similarities and differences between McClintock and Goldschmidt on the nature of genes (Keller 1983). The Biology and Gender Study Group (1989) essay extends the discussion to two other models of nucleus-cytoplasm relationships, which in turn are tentatively suggested to be related to cultural differences in family structures of the scientists who produced the theories. See Sapp (1987) for a history of cytoplasmic inheritance research.

For another feminist critique of conventional accounts of DNA, which includes a good discussion of the marginalization of Franklin in the DNA saga, see Hubbard (1990). The quotation about Watson and Crick's idea of the place of a feminist is from the Biology and Gender Study Group essay (1989). On McClintock, see the biography by Keller (1983), which includes the well-known but controversial chapter that bears the book title, *A Feeling for the Organism*. McGrayne's biographical chapter on McClintock (McGrayne 1993) includes a discussion of the geneticist's reaction to the biography by Keller. See also Keller (1985, 1989a) for further discussions of the topic. On the question of feminine/feminist method, see the critique of the idea of a universal feminist method from an SSK perspective by Richards and Schuster (1989a), with a reply by Keller (1989b) and counterreply by Richards and Schuster (1989b). For the general argument against a single feminist methodology, see the essays by Harding and Longino in Tuana (1989). For the extension of the argument to other underrepresented groups, see the concluding sections of Haraway's "Bio-Politics of a Multicultural Field" (1989:chap. 10). See Reinharz (1992:145) for sources on Ida Wells as well as for a general survey of the wide variety of research methods used in feminist studies.

For additional criticisms of racism and the concept of race in biology and anthropology, see especially the essays collected in Harding (1993), which includes a somewhat shortened version of Haraway's "Bio-Politics" essay. For criticisms of the nineteenth-century brain studies and differential intelligence, see Gould (1981) and Lewontin, Rose, and Kamin (1984:chap. 6). On the intelligence testing debates, see Gould (1981), Richardson (1990), and Snyderman and Rathman (1988). On Turing and the homosexuality trial, see Hodges (1983); also see Caporael (1986) on Turing and intelligence. On psychiatric/neurological theories of women's inferiority, see Braude (1989) and Brumberg (1988), among many other studies. Another version of the ongoing biological construction of inferiority involves the more complicated research on hemispheric lateralization and gender (Genova 1989; Rosser 1992:72–76). Also, watch for forthcoming work by Joe Dumit on PET scans and cultural difference (forthcoming in Downey, Dumit, and Traweek n.d.). For a more detailed comparison of French and Anglo-Saxon scientific styles, see Nye (1993) and Duhem (1982:chap. 4).

On the bricoleurlike or tinkerer quality of laboratory work, see Latour and Woolgar (1979) and Knorr-Cetina (1981). In terms of the temporal cultures discussed in chapter 4, *bricolage* is located somewhere betwixt and between modernist and postmodern anthropology. As a modernist concept bricolage implies a notion of systems, of sets of elements that pass from one discrete system to another, where they are reorganized and reconfigured as a new system (and therefore cast into a new kind of equilibrium—the new myth or reconstruction). Given the associations between the term *bricolage* and the modernist aspects of Lévi-Strauss's structuralism, bricolage is likely to be suspect among those who more squarely identify themselves with postmodern theorizing. Yet, I would suggest that bricolage is also a postmodern category, for it is concerned with transgressing boundaries, with disruptions of notions of closed and discrete systems, and ultimately (as in Lévi-Strauss's multivolume study of myths) with a lack of closure. One might also think of bricolage in terms of the rejection of science and technology. In addition to obvious cases such as groups or nations that reject sciences for religious or political reasons (e.g., Christian Scientists, the former Soviet Union), a well-known case is Tokugawa Japan's rejection of the gun (Perrin 1979). However, the Japan gun case is complicated, and I have not used it here because of what appear to be serious methodological problems (see Totman 1980).

The discussion of reconstruction does not explore class as a variable, in part because that issue is covered in chapter 6 in discussions of reconstruction by workers and by threatened local communities. See Dickson (1984) and Sclove (1993) for ways into the literature, especially ideas for democratizing technology policy and production.

The discussion of man-the-hunter vs. woman-the-gatherer draws on Haraway (1989:chaps. 7, 8, 11, 14; 1991:chaps. 2, 5) and Lee (1984) on hunter-gatherer calorie usage (see also Asquith 1981, 1986). For a historical discussion of gender, culture, and science, see Jordanova (1989). On the Ford probe, see the article by Vandermolen (1992). On a feminist approach to system design, see Benston (1989).

Regarding the reconstruction of primatology across cultures, see Haraway's "Bio-Politics" essay again (1989:chap. 10). That chapter would be my candidate for the best way in to the work of this influential STS/feminist/cultural studies historian.

For more information on George Washington Carver, and for the biographical sources consulted in this section, see Elliott (1966), Holt (1943), Manber (1967), McMurray (1981), and Winston (1971). On the use of peanut milk in Africa, see Manber (1967:121). The biographers also mention the bulletins written by Carver and available as *Bulletins, the Experiment Station*, nos. 1–44, Tuskegee Normal and Industrial Institute, 1898–1943. There is little information on the work of African or African American women scientists that could be compared to the culturally rooted project of Carver. See Rosser (1992:16–21) for a starting point regarding African American women biologists. It may be that the research simply has not been completed yet, or the lack of research may be the product of the large-scale exclusion of African and African American women from the scientific enterprise. However, there are some interesting cases of African American women inventors, including some inventions in the areas of cosmetics and domestic technology that are rooted in the experience of African American women (see Stanley 1983, 1993).

On the actor-network theories, again see Latour (1987) and Callon, Law, and Rip (1986). In the actor-network theory "things" such as machines, facts, and organizations can have effects on the growth of networks to such an extent that things can even be thought of as having some properties similar to actors. The theoretical differences in STS between, say, the ethnomethodologists of Britain and the United States, who are relatively individualistic, and the actor-network theorists of France, whom I would argue are Cartesian and holistic, strikes me as an example of the reproduction of French and Anglo-Saxon cultural styles. On strategies of circumvention, see Hess (1992a).

3. The Origins of Western Science: Technototems in the Scientific Revolution

For an overview of ancient science, see Lloyd (1970, 1973). Some classic examples of stories of the scientific revolution are Butterfield's *The Origins of Modern Science* (1957), which could be described as whig history but has the advantage of reaching back into the period before 1500; Hall's *The Revolution in Science, 1500–1750*, a careful scholarly volume that has gone through many editions without changing the basic Copernicus-to-Newton plot; and Bernal's *Science in History*, vol. 2 (1969), which takes a Marxist perspective. For a good survey of recent scholarship on the scientific revolution, see *Reappraisals of the Scientific Revolution* (Lindberg and Westman 1990), a volume that nevertheless does not examine in detail the question of non-Western influences. The introductory chapter by Lindberg (1990) contains a good survey of the history of the history of the scientific revolution.

The multivolume series on science and technology in China, by the historian and sinologist Joseph Needham, describes a number of factors that could have inhibited the emergence of modern science in China. In one summary statement of his work Needham offered an explanation that rested on the types of societies in the two areas. The feudal society of China was a bureaucratic type, whereas European feudal society was an "aristocratic military" type (1974:103). Needham argued that the Western type of feudal society was a "major factor" behind the emergence of "mercantile and then industrial capitalism, together with the Renaissance and the Reformation" (1974:103). See Restivo (1979) for a summary and critique of the various ways in which Needham explained why the "scientific revolution" did not occur in China. Although Needham buys into some of the assumptions of the standard version of the story of scientific revolution, he also sees his work as celebrating non-Western achievements (Needham 1973).

For the history of the history of Arab influences on Copernicus, see Saliba (1991). In addition to Ibn Shátir historians have suggested that Copernicus also borrowed from or was influenced by the Arab astronomers Nasir al-Din al-Tusi (Joseph 1991:347), al-Battání (Neugebauer 1969:206), and Thábit bin Qurra (Roberts 1957:432). Apparently the Copernican theory was really not simpler than the Ptolemaic theory. The historian O. Neugebauer notes that "the Copernican models themselves require about twice as many circles as the Ptolemaic models and are far less elegant and adaptable" (1969:204). Likewise, Saliba refutes Alexandre Koyre's claim that Copernicus abandoned the equant (Saliba 1991:76). Apparently, Copernicus is on the decline and Ibn al-Shátir is on the rise. For a review of the scholarship on Arabic-Moslem science and its influence on the modern West, see Huff (1993).

Ronan's *Science: Its History and Development Among the World's Cultures* (1982) does a fine job of surveying science in non-Western cultures, although it does not examine in detail the role of non-Western sources in the story of the scientific revolution. Pacey's *Technology in World Civilization* (1990) provides a similar survey, for technology rather than science, that is both accessible and concerned with interchanges between non-Western and Western cultures. Restivo's *Mathematics in Society and History* (1992) is the first book-length sociological account of the history of mathematics, and the book includes studies of non-Western mathematical traditions and their social bases. Although Joseph's *The Crest of the Peacock* (1991) provides a good survey of a multicultural approach to the history of mathematics, readers who consult that book may wish to consult Pingree (1993) for a discussion of some possible errors of source interpretation. On Hindu/South Asian roots of Western knowledge, see Goonatilake (1992b), and for a general critique of the ethnocentrism of historiography on the scientific revolution, see Rashed (1980).

See Merton (1970) and Ben-David (1971) for sociological accounts of the institutionalization of science in the seventeenth century, and on autonomy, see Karp and Restivo (1974:140). The problem of understanding and defining modernity has passed from Weber to Foucault, and the similarities and differences between the two influential social theorists invites comparison (see Gordon 1987). On the relationship between social and natural philosophy in the early modern period, see the suggestive discussions of Hobbes and modern science/society by Bajaj (1988), Shapin and Schaffer (1985), and Visvanathan (1988). My argument emphasizes the structural parallel between scientific and political concepts such as natural/social laws or methodology/constitutions (see Bloor 1982). On Protestantism and justification by faith, see Parsons (1968:53–54).

The Merton thesis has been the topic of a great deal of discussion and debate in the historical literature. Hill developed an alternative interpretation that linked Puritanism to science via an increase in antiauthoritarian attitudes and radical thinking in general (Hill 1985), and he also saw both Puritanism and science as outcomes of the more general shift from an agrarian to an urban/industrial society (Hill 1964). Hill's alternative formulation is only one among many; the Merton thesis has been attacked from virtually every angle possible, yet the experts still seem to recognize a general connection among Protestantism, capitalism, and the emergence of modern science. For reviews of the Merton thesis and the scholarship on it, see Abraham (1983), Lindberg and Numbers (1986), Shapin (1988), and Webster (1986).

Although I have used Margaret Jacob's survey, *The Cultural Meaning of the Scientific Revolution* (1988), that book builds on her earlier work (especially 1976) and that of James Jacob (1977). For the discussion of Giordono Bruno, Galileo, and Newton, see Lerner and Gosselin (1986) and Yates (1964; cf. Westman and McGuire 1977). See also the Lindberg and Westman (1990) volume, which includes a reevaluation of Margaret Jacob's work on atheism and pantheism (Hunter 1990) and an article that surveys the debate on the Yates's thesis and the influence of hermetic and occultist ideas on Newton (Copenhaver 1990). Historians who have studied the question of hermeticism and Newton subsequent to Yates's work of the 1960s and 1970s have tended toward the view that the general tradition of alchemical and magical writings influenced Newton, but the specific tradition stemming from Hermes Trismegistus had only a minor influence. See also Hill (1975), who situates Newton in seventeenth-century society. On the Puritans' attack on magic, see Thomas (1971), and on the connections between witchcraft and capitalism, see Boyer and Nissenbaum (1974) and McFarland (1970). On the comparison of radical Puritan groups to communists, see Walzer (1965). On the natural philosophers and the constructions of witchcraft as pathology, see Merchant (1980:140–43). For a general discussion of the argument that nature is to culture as female is to male, see Ortner (1974) and MacCormack and Strathern (1980). On New Age/skeptic debates from the nineteenth century to the present, see Hess (1993).

For an extension of Merchant's argument to the case of Robert Boyle in the context of a critique of Deason, see Keller (1992:62–67). On the scientific expeditions to the Amazon, see Hecht and Cockburn (1989:6–8). On the Dutch in Brazil, see Boxer (1957:chap. 4), and on the British use of botanical gardens, see Brockway (1979). On biotechnology and neocolonialism, see Busch (1991).

4. Temporal Cultures and Technoscience

The concept of paradigms and epistemes is foreshadowed in the earlier work of Ludwik Fleck (1979), whose book on thought styles and thought communities is now recognized as a classic that prefigured the social constructivist revolution in the sociology of science. A number of introductions to Foucault's work is now available; readers may wish to begin with Dreyfus and Rabinow (1982). Stemerding (1991) compares Foucault's work to a Latourian analysis of changes in taxonomy, and he shows how the two approaches are complementary. I would add that actor-network analysis becomes unwieldy as an explanation of cross-disciplinary parallel changes (that is, epistemic shifts rather than disciplinary changes), as it does for other kinds of macrostructural and cultural questions. Furthermore, the focus on networks rather than structure tends to lead to a gradualist view of history. On the gradualists versus revolutionists in the history of science, see Lindberg (1990).

See Dumont (1977, 1986) for his analysis of individualism. Readers interested in connecting Dumontian analysis with the natural sciences may wish to consult Holton's discussion of the discreteness thema, which he argued includes the theory of atoms proposed by Dalton in 1808, the theory of cells proposed by Schleiden for plants in 1838 and by Schwann for animals in 1839, the idea of genes proposed by Mendel in 1865, Joule's theory that heat results from the motion of atoms and molecules (proposed in 1847), the discovery of the electron in 1897, and the quantum theory of radiation proposed in 1900 and 1905 (Holton 1988:84). Foucault has criticized the concept of individualism as nonspecific (1986:42–43), and he has also distinguished his epistemic analysis from more general forms of cultural analysis (1991). However, I think that a general analysis of temporal cultures can be useful if it is not reduced to the templates against which Foucault warned. For the more empirically minded who are not comfortable with the holistic and global perspective that I have adopted here, Forman's (1971) analysis of Weimar culture and physics provides another example of an analysis of science in temporal culture, and one that is much more specific in terms of its restriction of time and place and its methodological precision (although not without controversy among historians; cf. Hendry 1980).

On technology in the transitions among temporal cultures, Jameson (1984) follows Mandel (1975:118–119) in linking cultural logics and social structures to transitions in the machine production of different types of motors: steam since 1848, electric and combustion motors since the 1890s, and electronic and nuclear-powered apparatuses since the 1940s. Because I see technology as material culture, I am less willing to see the machine production of types of motors as anything more than one marker of fuzzy historical transitions. Furthermore, the cultural perspective that I have taken has expanded the question of technology to include the appearance of, for example, electric systems or the computer.

On entropy in comparison with evolution (as well as the twentieth-century articulations via discussions of dissipative structures), see Porush (1992), and on the general cultural interpretations of entropy in the nineteenth century, see Myers (1989). On embryology, including its transition from a descriptive to experimental science in the late nineteenth century, see Gilbert (1991). Modernist human and administrative sciences in France are covered well by Rabinow (1989). On equilibrium theory in modernist economics, see Marshall's *Principles of Economics*, book 5 (1961). Heilbroner (1972:201) explains that the concept of equilibrium existed in prior economic theory, but Marshall refined the concept and brought it to the center of what is today called microeconomic theory. On equilibrium concepts in early twentieth-century biology, see Allen (1978:95–111, 126–45).

Rather than claim that notions of dynamic equilibrium and closed systems are unique to the modernist period, I am claiming that they are characteristic of the period. Thus, versions of equilibrium theory appear before and after the modernist period. For example, the physiological theories of homeostasis of Cannon and Henderson can be traced back to Claude Bernard (Cross and Albury 1987), and Canguilheim (1988:chap. 4) has traced the idea of regulation even further back into the nineteenth century. In physics the ideas of thermal and mechanical equilibrium attest to the long history of that concept, and in political theory Gulick (1955:156) shows that the concept of equilibrium was explicitly in use in the early nineteenth-century system of balance-of-power politics. I am concerned here with a specific form that involved dynamic equilibrium in closed systems and that is elaborated in a number of disciplines during the modernist period.

For a historical background on psychology, see Ellenberger (1970) and Leahey (1992). Pavlov also described the "analysis of equilibration" as the "primary aim of physiological enquiry" (1928:49). Freud's structural model appears in *The Ego and the Id* (1962) and is elaborated in his subsequent work as well as in the defense mechanism models of his students. Equilibrium concepts are most evident when Freud is using his economic metaphors. I have not discussed relativity theory or the uncertainty principle because they would seem to suggest the beginnings of postmodern science in that they are concerned with observer/observed relations. However, I leave this as an open question. For an example of equilibrium theory in early twentieth-century chemistry, see Lewis and Randall's *Thermodynamics* (1961), originally published in 1923.

For a useful introduction to Saussure in the context of the human sciences, see Culler (1976). The transition from literary history to the New Criticism in literary studies that occurred during the mid-twentieth century is another example of the displacement of historical and evolutionary approaches in the human sciences, and New Criticism has elsewhere been compared to functionalism in the social sciences (Boon 1982).

On Marshall, Pareto, and Parsons, see Parsons's *The Structure of Social Action* (1968). Henderson is said to have influenced Parsons, Homans, Kluckhohn, and Brinton through the Harvard Society of Fellows and his Pareto seminar. That issue as well as Cannon's and Henderson's application of the homeostasis model to social problems is discussed by Cross and Albury (1987). Parsons was also one of a number of social theorists who applied cybernetic theory to the social world (see Parsons 1977; Heims 1991). As Heims points out, cybernetics influenced a number of twentieth-century social scientists in addition to Parsons, most prominently the polymath (and prepostmodernist!) Gregory Bateson. Finally, on structural anthropology—which provides an alternative to functionalism, but often with modernist notions of formal systematicity—see Lévi-Strauss (1963a) for an overview, Lévi-Strauss (1969) on kinship, and Lévi-Strauss (1975) for the first volume on myth.

On chaos and complexity theory I have gone little beyond the popular accounts of Gleick (1987) and Waldrop (1992). Gleick's book on chaos reveals a more sophisticated social understanding of science and technology than Waldrop's book on complexity, and Kellert (1993) introduces the topic through the lens of STS research. See the critical review of Waldrop by Winner (1993a), whose skepticism I support and echo. On chaos/complexity theory and evolution, see Kauffman (1993). For a discussion of boundary transgressions and cyberspace, see Stone (1992), and on boundary transgressions in physics and anthropology, see Traweek (1992a). The shift implied by molecular biology and the genome project has received attention from a number of perspectives. In addition to Rabinow (1992a, 1992b), see Goonatilake (1992a), who conceptualizes the transformation in terms of a merging of the evolution of genes and culture. Busch (1991) views it in terms of a shift in the longstanding tradition of plant improvement, but one in which plants have now become manufactured items. See Rothstein and Blim (1991) for anthropological perspectives on the postmodern economy.

On the new ethnographies, see Clifford and Marcus (1986), and on reflexive STS studies, see Woolgar (1988b). On feminism, positionality, and ethnography, see Gordon (1988), Kirby (1991), Mascia-Lees, Sharpe, and Cohen (1989, 1991), Traweek (1992a), as well as general discussions by Haraway (1991). See Christopher Johnson (1993:151) for a discussion of the parallel between Derrida's deconstruction and open-systems models of cybernetics, and McHale (1987) on postmodern literature. For examples of New Age uses of science, see Ferguson (1980) and Talbot (1991). On the critique of New Age theories in addition to Hess (1993), see Restivo (1985) for a full critique of the parallelist arguments. For examples of chaos/complexity theory that shows how it is being applied to society, see DeLanda (1992) and Lewin (1992). See also the review essay on cyberspace and anthropology by Escobar (1994).

5. The Social Relations and Structures of Scientific and Technical Communities

My discussion of transnational communities and diasporas benefited from the 1993 annual meeting of the Society for Cultural Anthropology, where Arjun Appadurai, James Clifford, David Edwards, Faye Ginsburg, Barbara Kirshenblatt-Gimblett, Margaret Mills, Roger Rouse, and others gave lucid presentations. Those presentations were scheduled to appear in a special issue of *Cultural Anthropology* in 1994. For a way in to the literature on cultural borders, see the journal *Diasporas*.

Dumont's concept of hierarchy is not without its critics, and the term *hierarchy* in *Homo Hierarchicus* may have as many meanings as the term *paradigm* in Kuhn's *Structure of Scientific Revolutions*. See Appadurai (1992) for an intellectual genealogy of Dumont's ideas and Kolenda (1976) on Dumont's uses of the term *hierarchy*.

I have focused on intercultural communication as an alternative to other types of acultural approaches to communication in STS, such as the rhetoric of science, ethnomethodological conversational analyses, and discourse analysis. Again, my goal is to help build a vision of cultural studies of science and technology that puts culture/power issues back on the center stage. For introductions to intercultural communication, see Porter and Samovar (1991), Hall and Hall (1990), and the *International Journal of Intercultural Relations*. For a more research-oriented rather than popular view of gender and communication, see Tannen (1993). The continuum of cultures that I listed comes from Porter and Samovar (1991:20). On the comparative study of touching, see Argyle (1975:290); on Colombian handshakes, see Gorden (1987:203); and on Latin Americans and furniture etiquette, see Morain (1987:127). The examples of Japanese-American cultural misunderstandings given by Hall and Hall (1987) may have been fictionalized or composite accounts; I have summarized them because they provide good examples of what kinds of problems American and Japanese are likely to encounter when they do business with each other. I have developed some of the insights of Kant de Lima and DaMatta in my ethnographic analyses of Brazilian Spiritist conferences (Hess 1991:chap. 2).

Figure 5.2 is compiled from the *AAA Guide 1992–93*, the guide to departments of anthropology published by the American Anthropological Association (Washington, D.C.). On departmental hiring patterns in anthropology, see Bair, Thompson, and Hickey (1986). On the tendency for what I would call virtual endogamy (departments hiring their own members), see Gross (1970). Rankings vary quite a bit over time and by the type of criteria used; see, for example, Gourman (1989:11) versus Bair, Thompson, and Hickey (1986). On Washburn's lineage in primatology, see Haraway (1989:chap. 8, 1991:chap. 5).

For national styles, Harwood's *Styles of Scientific Thought* (1993) is a good way in to the literature. My discussion of Harwood relies on his 1987 article in *Isis* as well. Other frequently cited sources on the topic include Reingold (1978) and

Schweber (1986) on physics. Nye (1993) and Geisson (1993) consider the question of comparative scientific styles in the context of comparative studies of research schools. The classical STS studies on comparative university structures are by Ben-David (1971:chaps. 7–8, 1991:chaps. 5–10); however, those studies focus on European and American universities, the nineteenth and early twentieth century, and general social function rather than the interrelationship of social structure and style. See also Ringer (1979) for another comparative discussion of higher education in Europe that does not focus on departmental structure issues. Discussions of Japanese social structure and hierarchy from a comparative perspective often begin with Benedict (1946); I am grateful to Sharon Traweek for pointing me to the much more subtle analysis by Nakane (1970).

For an overview and history of science in Brazil, see Schwartzman (1991) and Stepan (1976) and, in Japan, see Bartholomew (1989). On science in Latin America, see the journal *Quipu: Revista Latinoamericana de Historia de las Ciencias y la Technologia* (Mexico City). For a more specific discussion of intercultural communication issues and Brazil, see Harrison (1983). For a general exploration of hierarchy and personalism in Brazil by a number of scholars, see Hess and DaMatta (1994). See also Botelho (1989) on Brazilian scientists and the military dictatorship. The hierarchy and personalism of Latin America can provide a context from which to read STS actor-network theories as culturally located constructions; as Botelho (1990) argues, the model does not work very well for his Brazilian materials. However, at the same time France shares some structures of hierarchy and personalism with the Iberian and Latin American cultures, and as a result network-oriented theories may make more intuitive sense there than in more individualistic cultures such as the United States.

I have not discussed scientific communities in other nations because the chapter seemed already long enough, and the four cases set up ways of approaching other cultural differences. As for the case of Latin America, Indian and African universities are generally a legacy of the colonial powers, but there is little research on how indigenous notions of social structure have reshaped the colonial institutions. There are some intriguing suggestions in the literature, such as D. N. Dhanagare, who notes, "Whereas in the British universities authority has devolved from the professor or head of each department to the departmental academic staff as a whole, in the Indian universities it has in general remained as it used to be" (1984:396; see also Béteille 1981). There is a substantial literature on laboratories and scientific communities in India, e.g., Anderson (1978), Visvanathan (1985), Shiva and Bandyopadhay (1980), and the collection "Reports and Documents: Science in the Indian Universities," which appeared in the 1992 issue of *Minerva* 30(1):51–100. On efforts to build scientific communities in the third world, see Segal (1987). For a readable, journalistic account of the special conditions of doing science in Africa, see Bass (1990). The journal *Minerva* frequently carries articles on science in the third world.

6. Science and Technology at Large: Cultural Reconstruction in the Broader Society

Zabusky (1994) has a book forthcoming on the topic, which promises to explore the issue of scientists in the ESA in more detail. My general characterization of the diversity of the movement for environmental justice comes from Gelobter (1993). Deerinwater is also a cofounder of Citizens United to Rescue the Environment (CURE), an umbrella group for antinuclear organizations that is an example of the coalition politics that have emerged out of the environmental justice movement (Deerinwater and Smith 1993). On the double injustice of women and environmental injustice, see Chavkin (1984), Mullings (1984) on women of color in the U.S., and Shiva (1989) and Dankelman and Davidson (1988) on women and the environment in the third world. My student Kenneth Croes (1994) has also found a similar strategy in which the Cree and the environmentalists use official data. However, the local coalition has also recruited outside scientists, who have generated new research. Also watch for forthcoming work by Kim Laughlin on the Bhopal disaster, gender, culture, and environmental politics.

On creationism and science, see Nelkin (1982) and Toumey (1991). For a survey of positions both for and against New Age thought in the United States, see my book *Science in the New Age* (1993). The book discusses the question of "boundary work" (Gieryn 1983) between science and religion as well as complex boundaries drawn among and within groups generally dismissed as nonscientific. Andrade's Spiritist science is summarized in a three-volume magnum opus and elaborated in a number of other publications; for an introduction, see Andrade (1986). My own work on Spiritism is summarized in Hess (1991, 1994b). A slightly more detailed summary of my discussion of Andrade's paraphysical theories is in Hess (1994a).

For the research on amniocentesis, see Rapp (1988, 1990, 1991) and an essay forthcoming in Downey, Dumit, and Traweek (n.d.). The field of anthropological and cultural studies of reproductive technologies is growing rapidly. Among the significant studies not discussed here are Clarke (n.d.), Clarke and Montini (1993), Edwards, Franklin et al. (1993), Strathern (1992), Shore (1992), and the collection forthcoming by Rapp and Ginsburg (n.d.).

My discussion of women and technology focused on women's reproductive technologies, but there are also a number of other sites for the exploration of women's perspectives on technology. One area involves women's responses to changing technologies in the workplace (e.g., Barker and Downing 1985, Cockburn 1985). The emergence of industrial technology in the home appears not to have reduced women's domestic labor. Instead, it has been accompanied by the "proletarianization" of the housewife because the increased domestic technology was accompanied by an increase in the responsibilities of the housewife and a loss of domestic servants in middle-class homes (Cowan 1985b). Some women's groups have advocated and experimented with revolutionary new domestic structures, such as centralized kitchens (Doorly 1985).

On the non-Western forms of worker resistance, see Ong (1987) on Malaysian spirit possession, and Taussig (1980) and Nash (1979) on Andean miners. On flexible systems production and contemporary global production, see Harvey (1989). The statistics on Japanese plants in the U.S. come from Fucini and Fucini (1990:219). On other Japanese automobile factories, see Gelsanliter (1990). My discussion of Harvey, flexible systems production, and total quality management has benefited from conversations with Emily Martin during a visit to Rensselaer. She has been exploring these developments from a cultural viewpoint. On the new Taylorism and the appropriation of anthropology in corporate culture, see Martin (1992) and her forthcoming book (1994).

7. Other Ways of Knowing and Doing: The Ethnoknowledges and Non-Western Medicines

For a history and survey of ethnobotany, see Ford (1978). On the cassowary, see Bulmer (1967). On similarities among ethnobiological classification systems, see Berlin, Breedlove, and Raven (1973) and Boster, Berlin, and O'Neill (1986). On the question of the structured world versus structured mind theory of the similarities, see Boster and D'Andrade (1989). On similarities in color symbolism, see Berlin and Kay (1969). On Boas and the words for snow, see Martin (1986). For a succinct survey of the Mayan calendar and Mayan ethnoastronomy of Venus, plus detailed bibliographic references, see Aveni (1989), and on ethnoastronomy in the Pacific, see Goodenough (1953) or Goodenough and Thomas (1987). For a broader survey of New World tropical ethnoastronomy, see Aveni and Urton (1982). For a review of archaeoastronomy and ethnoastronomy, see Baity (1973), and for ethnopharmacology, see Etkin (1988). For ethnoknowledges in general there is an encyclopedia forthcoming from Hampshire College librarian Helaine Selin; see also Turnbull and Watson-Verran (n.d.).

For an introduction to and survey of ethnomathematics, see Ascher (1991). The list of ethnomathematical topics is drawn from her book. For discussions of ethnomathematics from the point of view of pedagogy and work that draws on Paulo Freire's (1970) pedagogical theories, see Borba (1990, 1992) and D'Ambrosio (1985), and on the Australian project, Watson-Verran (1988). On U.S. shoppers, with a good general comparative discussion, see Lave (1988), and on South African carpenters, see Millroy (1990, cited in Borba 1992). For a historical survey of non-Western Old World mathematics, see Joseph (1991).

On intellectual property rights, see Posey (1990) and the special issue of *Cultural Survival Quarterly* (1991), vol. 15, no.3, devoted to the topic. That issue includes a description of the Kuna guidelines (Chapin 1991), a survey of the work of international organizations (Posey 1991), and discussions of related Native property rights issues for songs, textiles, and artifacts. On indigenous knowledge and development, see the collection edited by Brokensha, Warren, and Werner (1980), including the essay on intercropping techniques by Belshaw (1980). See also the *Proceedings of the First International Congress of Ethnobiology* (Posey and Overal 1990).

On the efficacy of non-Western healers, see the classic "Effectiveness of Symbols" in Lévi-Strauss (1963a) and "A Ndembu Doctor in Practice" in Turner (1967). Also, see the case studies in Crapanzano and Garrison (1977). On medical pluralism, see Leslie (1980), and on medical domination and hegemony, see Baer (1989), Elling (1981), Frankenburg (1980), as well as Taussig (1987). The essay cited by Apffel Marglin (1990) draws on a large literature on Indian variolation and has a bibliography for those interested in following up on the topic in more detail. For the *New England Journal of Medicine* article on the use of alternative medicine in the United States, see Eisenberg et al. (1993). Although non-Western therapies were included on the list, the four top therapies of choice were chiropractic, relaxation, exercise, and prayer. Those categories are ambiguous enough that they may have included some non-Western practices. On African/Latin American therapies in the U.S., see A. Harwood (1987).

For a review of some related alternative psychotherapies from Japan, see Reynolds (1980) on the Japanese "quiet" psychotherapies. For a critique of the endorphin release hypothesis for acupuncture, see Skrabanek (1989:190–91). For a review of Becker's acupuncture research, see Becker and Marino (1982:197–99). Ainslie Meares's research and related therapies are reviewed in the Office of Technology Assessment's *Unconventional Cancer Treatments* (1990). Related to the meditation research are the claims that imagery may lead to regression of cancer and related diseases. The research is very controversial and by no means settled at this time. For an example of the methodological problems that have been raised, see Friedlander (1989).

Regarding the illness/disease dichotomy, see Frankenburg (1980) and Young (1982). Hahn proposes a tripartite system like that of Frankenberg and Young, but one in which the third term is not sickness but instead disorder: the construction of suffering by traditional or non-Western healers (Hahn 1984:16ff.). In contrast, I have used the word *disorder* for psychological/psychiatric conditions rather than the word *disease*, which I reserve for biomedical conditions. Hahn also argues that his framework is "pluralistic and ecumenical" (16), unlike that of Kleinman and his critics: "While Frankenberg and Young have corrected a view of disease and illness which neglects the powers of society, they, like Kleinman . . . have confused an account of the phenomena with the phenomenon accounted for" (14). Keyes (1985:167), however, argues that Kleinman escapes from this confusion in a subsequent definition of illness and disease that emphasizes that the latter is also the practitioner's construction of the patient's ailment (see also Kleinman 1988:25, 89–90; Obeyesekere 1985:136). A subsequent collection in psychiatric anthropology is very clear about the constructed nature of both local and cosmopolitan medicines; the title reads *Ethnopsychia-*

try: The Cultural Construction of Professional and Folk Psychiatries (Gaines 1992).

On the politics of AIDS research there is a vast literature, but my candidate for a sophisticated starting place at the nexus of STS and cultural studies is Treichler (1991). On the history of the construction of Western psychiatric disorders, see Bernheimer and Kahane (1985) on hysteria and Veith (1965) for a history of hysteria that provides some discussion of hysteria in men, including during the pre-Freudian era. See also Abelove (1993) on Freud and homosexuality, Jackson (1985) as well as other essays in Kleinman and Good (1985) on depression, and Kenny (1986) and Rosenbaum (1980) on multiple personality. On comparative aspects of mental illness, see Kleinman (1985) on neurasthenia in China, Prince (1964) on traditional Yoruba psychiatry, Simons and Hughes (1985) and Prince and Tcheng-Laroche (1987) on culture-bound syndromes, and Brumberg (1988) and Di Nicola (1990a, 1990b) on anorexia.

8. Cosmopolitan Technologies, Native Peoples, and Resistance Struggles

For the basic texts of the twentieth-century American evolutionary anthropologists, see White (1949, 1987), Steward (1955), and Sahlins and Service (1988). See also Harris's discussion of White, Steward, and related figures as well as a defense of Harris's own "cultural materialism" in Harris (1968). On interpretive flexibility, see the Empirical Program of Relativism (Collins 1985) and a review of the "turn to technology" in STS (Woolgar 1991).

For a general survey of the impact of development projects on indigenous peoples and some of their responses, see Burger (1987) and, for the Chipko movement, Shiva (1989:67–77). On the appropriation of Native territories for nuclear testing, see Dibblin (1988), Nietschmann and LeBon (1987), and Kiste (1987). The U.S. testing in the South Pacific is only one example of Native peoples from all over the world who have suffered from nuclear testing. See also Gusterson (n.d.) on nuclear weapons scientists. On the Alaska North Slope project I also consulted Condon (1989). On the impact of hydroelectric projects, see the special issue of *Cultural Survival Quarterly* (1988), vol. 12, no. 2, as well as Chernela (1988) and Lawson (1982).

For a general discussion of ethnographic film, indigenous media, and the complexities of cultural mediation, see Ginsburg (1992, 1994). On the Kayapó and media, see Turner (1991, 1992) and the films *The Kayapó of Gorotire* and *The Kayapó: Out of the Forest*, both produced and directed by Michael Beckham (*Disappearing World Series* of Grenada T.V., respectively 1987 and 1989). On the Inuit, see *Magic in the Sky,* directed by Peter Raymont (National Film Board of Canada, 1991). See also discussions in the journal *Visual Anthropology*.

On the Iroquois, the Kinzua Dam, and other incursions on their territory, see Hauptman (1986). On the civil engineer and utopianist Arthur Morgan, see Talbert (1987), and on the public works mogul Robert Moses, see Caro (1974). For

Hydro-Quebec there has been regular coverage in the *New York Times* and Albany *Times Union* as well as in some magazines. For a detailed discussion of Hydro-Quebec in the context of New York State environmental politics, see the master's thesis by my student Kenneth Croes (1994).

On the idea of technological fix, see Ellison (1978). For the discussion of the political background of the Green Revolution, I have relied on Anderson and Morrison (1982) and Anderson (1991), with some support from DeWalt and Barkin (1987) on Mexico. On the relationship between gender and increased poverty, see Lipton and Longhurst (1989). See Shiva (1989) for a general discussion of feminist and ecological issues related to the Green Revolution, and Nanda (1991) for a critique of Shiva that discusses the issues of amniocentesis and wage/subsistence farming. On the loss of genetic diversity, see the educational film *Seeds of Tomorrow* (produced by WGBH for the *Nova* series, 1985). On the Shri AMM Murugappa Chettiar Research Centre and biodynamic gardening, see Seshagiri and Chitra (1983). On nonhybrid seeds and the ability to combine the Green Revolution with local and traditional agricultural practices, see Nanda (1991).

For Vicos, see Holmberg et al. (1965), Doughty (1987), and, on gender inequities, Babb (1985). For a study of the difficulties of the coalition politics involved in the demarcation of lands and biospheres, see also Chernela (1990), Ehrenreich (1990), and Jahnige (1989). For a survey of feminist perspectives on development ideologies, see Warren and Bourque (1991). For an introduction to the women and development literature, see Tinker (1990).

For a survey of grassroots development projects, see "Grassroots Economic Development," a special issue of *Cultural Survival Quarterly* (1987), vol. 11, no. 1. That issue includes abstracts of and references to the projects cited among the Irulu, Huichol, Ladakhis, and Shuar. For additional examples in the context of the appropriate technology movement, see Pacey (1983:chap. 8, 1990:chap. 11). On the background of the Trombe Wall and Ladakh Project, now the International Society for Ecology and Culture, see Norberg-Hodge (1991). The project of building local ponds for water is based on the video *We Can Solve It,* presented by Ashvin Shah at Rensselaer Polytechnic Institute for the conference "Non-Western Perspectives on the Environment: The Third Comparative Science and Technology Conference," April 1993. For a publications list of the Shri AMM Murugappa Chettiar Research Centre, write to that center at Tharamani, Madrass 600 113, India. See also the Indian journal *PPST Bulletin* (for People's Patriotic Science and Technology).

On the Australian Aboriginal media projects, see Fry and Willis (1989), Ginsburg (1992), and Michaels (1986). Native-controlled media such as the Aboriginal radio and television stations have the potential of being consumed by non-Native listeners, and as a result this technology could be classified as crossing the second and fourth categories. On Aboriginal art, see Benjamin (1990) and a more general problematizing of aboriginal art and its reception by Myers (1992). See also the journal *Studies in the Anthropology of Visual Communication.*

For general introductions to resistance movements, see Moore (1978), Piven and Cloward (1977), and Scott (1990). Scott (1985) also discusses resistance in the context of the Green Revolution among Malaysian peasants. Hill (1988) touches on the question of resistance in the context of technology, and Wright (1991) elaborates the idea of technologies of repression for national societies.

9. Conclusions: Science, Technology, and Multicultural Education

For the *Workforce 2000* Hudson Institute report, see Johnston and Packer (1987). The statistic is for the years 1985 to 2000. See Rosser (1992:30–31) for a survey of some of the statistics on women and underrepresented ethnic groups in the science and engineering professions. For example, although women in the U.S. scientific/engineering workforce grew from 9 percent in 1978 to 16 percent in 1988, only one in twenty-five women were engineers. Of the U.S. women scientists and engineers only 5 percent were African American, 5 percent were Asian American, 3 percent Latina, and less than 1 percent Native American. Asian American women are employed in the science/engineering professions at a higher percentage rate than in the general workforce. On the weeding-out process in engineering, see Downey, Lucena, and Hegg (n.d.). The 1992 Survey of Earned Doctorates for U.S. citizens from the National Research Council revealed that 19.7 percent of the doctorates in the physical sciences and 9.3 percent in engineering were for women. In the physical sciences 5 percent of the doctorates went to Asian Americans, 1 percent to African Americans, 2.5 percent to Hispanics, and 0.5 percent to Native Americans. In engineering 10 percent went to Asian Americans, 1.5 percent to African Americans, 2.7 percent to Hispanics, and 0.5 percent to Native Americans.

For some examples of work on women in science and engineering, see Abir-Am and Outram (1987), Carter and Kirkup (1990), and Kass-Simon and Farnes (1990). For research on women engineers and salaries, see McIlwee and Robinson (1992). On the STS curriculum and gender, see Heath, Glaser, Gudmendsen, Jansen, Lewish, and Rooney (1991). See also the policy recommendations of the Committee on Women in Science and Engineering of the Office of Scientific and Engineering Personnel of the National Research Council (1991). On the survey at the University of California, see Aufderheide (1992:182). Of the many consultants in the field Rensselaer benefited from a very helpful visit and workshop led by Reginald Wilson of the American Council on Education.

As I learned from attending the annual conference of the Fund for Improvement of Post-Secondary Eductation (FIPSE), there are many excellent FIPSE projects in this area, and the ideas are available in a booklet from the agency titled *Program Book: Project Descriptions*, compiled and edited by Sandra Newkirk and Susan McGraw. In addition to the Rensselaer and University of Cincinnati programs already mentioned, the booklet lists other relevant programs, including the Brown University Parallax program and Skidmore College capstone courses.

Bibliography

Abelove, Henry. 1993. "Freud, Male Homosexuality, and the Americans." In Henry Abelove, Michèle Aina Barale, and David Halperin, eds., *The Lesbian and Gay Studies Reader*. New York: Routledge.

Abir-Am, Pnina, and Dorinda Outram. 1987. *Uneasy Careers and Intimate Lives*. Rutgers: Rutgers University Press.

Abraham, Gary. 1983. "Misunderstanding the Merton Thesis." *Isis* 74:368–87.

Abu-Lughod, Lila. 1989. "Bedouins, Cassettes, and Technologies of Popular Culture." *Middle East Report* (July-August), pp. 7–11, 47.

Adas, Michael. 1989. *Machines as the Measure of Men*. Ithaca: Cornell University Press.

al-Hassan, Ahmad, and Donald Hill. 1986. *Islamic Technology: An Illustrated History*. Cambridge: Cambridge University Press and Paris: UNESCO.

Allen, Garland. 1978. *Life Sciences in the Twentieth Century*. Cambridge: Cambridge University Press.

Anderson, Robert S. 1978. *Building Scientific Institutions in India: Saha and Babha*. Montreal: McGill Centre for Developing-Area Studies.

— 1991. "The Origins of the International Rice Research Institute." *Minerva* 29(1):61–89.

Anderson, Robert S., and Barrie M. Morrison. 1982. "Introduction." In Robert S. Anderson et al., eds., *Science, Politics, and the Agricultural Revolution in Asia*. Boulder: Westview.

Andrade, Hernani Guimarães. 1986. *Psi Quântico. Uma Extensão dos Conceitos*

Quânticos e Atômicos à Idéia do Espírito. São Paulo: Pensamento.

Appadurai, Arjun. 1990. "Disjuncture and Difference in the Global Political Economy." *Public Culture* 2(2):1–24.

—— 1992. "Putting Hierarchy in Its Place." In George Marcus, ed., *Rereading Cultural Anthropology*. Durham: Duke University Press.

Arglye, Michael. 1975. *Bodily Communication*. New York: International Universities Presses.

Ascher, Marcia. 1991. *Ethnomathematics: A Multicultural View of Mathematical Ideas*. Pacific Grove, Cal.: Brooks/Cole.

Asquith, Pamela. 1981. "Some Aspects of Anthropomorphism in the Terminology and Philosophy Underlying Western and Japanese Studies of the Social Behavior on Non-Human Primates." Ph.D. diss., Oxford University.

—— 1986. "Anthropomorphism and the Japanese and Western Traditions in Primatology." In J. Else and P. Lee, eds., *Primate Ontogeny, Cognition, and Behavior*. New York: Academic Press.

Aufderheide, Patricia, ed. 1992. *Beyond P.C.: Toward a Politics of Understanding*. St. Paul: Graywolf Press.

Aveni, Anthony. 1989. *Empires of Time*. New York: Basic.

Aveni, Anthony, and Gary Urton. 1982. *Ethnoastronomy and Archaeoastronomy in the American Tropics*. New York: New York Academy of Sciences.

Babb, Florence E. 1985. "Women and Men in Vicos, Peru: A Case of Unequal Development." In W. Stein, ed., *Peruvian Contexts of Change*, pp. 163–209. New Brunswick: Transaction Books.

Bacon, Sir Francis. 1952. *Advancement of Learning. Novum Organum. New Atlantis*. Vol. 30. *Great Books of the Western World*. London, Chicago: Encyclopaedia Britannica.

Baer, Hans. 1989. "The American Dominative Medical System as a Reflection of Social Relations in the Larger Society." *Social Science and Medicine* 28(11): 1103–12.

Bair, Jeffrey, William Thompson, and Joseph Hickey. 1986. "The Academic Elite in American Anthropology: Linkages Among Top-Ranked Graduate Programs." *Current Anthropology* 27:410–12.

Baity, E. C. 1973. "Archaeoastronomy and Ethnoastronomy So Far." [Plus comments and reply.] *Current Anthropology* 14:389–449.

Bajaj, Jatinder. 1988. "Francis Bacon, the First Philosopher of Modern Science." In Ashis Nandy, ed., *Science, Hegemony, and Violence*. Delhi: Oxford University Press.

Barbosa, Lívia. 1994. "The Brazilian *Jeitinho*: An Exercise in Brazilian National Identity." In David Hess and Roberto DaMatta, eds., *The Brazilian Puzzle: Culture on the Borderlands of the Western World*. New York: Columbia University Press.

Barker, Jane, and Hazel Downing. 1985. "Word Processing and the Transformation of Patriarchal Relations of Control in the Office." In Donald MacKenzie and

Judy Wajcman, eds., *The Social Shaping of Technology*. Philadelphia: Open University Press.

Barley, Stephen. 1988. "On Technology, Time, and Social Order: Technically Induced Change in the Temporal Organization of Radiological Work." In Frank Dubinskas, ed., *Making Time: Ethnographic Studies of High-Technology Organizations*. Philadelphia: Temple University Press.

Barnes, Barry. 1981. "On the 'Hows' and 'Whys' of Cultural Change." *Social Studies of Science* 11:481–98.

Barnes, Barry, and Donald MacKenzie. 1979 ."On the Role of Interests in Scientific Change." In Roy Wallis, ed., *On the Margins of Science*. Sociological Review Monograph no. 27. Keele, Staffordshire: University of Keele.

Barnes, Barry, and Steve Shapin. 1979. *Natural Order: Historical Studies of Scientific Culture*. Beverly Hills: Sage.

Bartholomew, James. 1989. *The Formation of Science in Japan*. New Haven: Yale University Press.

Barzun, Jacques. 1958. *Darwin, Marx, Wagner*. New York: Anchor.

Bass, Sam. 1990. *Camping with the Prince and Other Tales of Science in Africa*. Boston: Houghton-Mifflin.

Becker, Robert, and Andrew Marino. 1982. *Electromagnetism and Life*. Albany: SUNY Press.

Bellah, Robert. 1957. *Tokugawa Japan*. Boston: Beacon.

Belshaw, Deryke. 1980. "Taking Indigenous Knowledge Seriously: The Case of Intercropping Techniques in East Africa." In David Brokensha, D. M. Warren, and Oswald Werner, eds., *Indigenous Knowledge Systems and Development*. Washington, D.C.: University Press of America.

Ben-David, Joseph. 1971. *The Scientist's Role in Society*. Englewood Cliffs, N.J.: Prentice-Hall.

— 1991. *Scientific Growth: Essays on the Social Organization and Ethos of Science*. Berkeley: University of California Press.

Benedict, Ruth. 1946. *The Chysanthemum and the Sword*. New York: New American Library.

Benjamin, Robert. 1990. "Aboriginal Art: Exploitation or Empowerment?" *Art in America* 78(7):73–81.

Benston, Margaret Lowe. 1989. "Feminism and System Design: Questions of Control." In Winnie Tom, ed., *The Effects of Feminist Approaches on Research Methodologies*. Waterloo, Ontario: Wilfrer Laurier University Press.

Berlin, Brent, and Paul Kay. 1969. *Basic Color Terms: Their Universality and Evolution*. Berkeley. University of California Press.

Berlin, Brent, Dennis Breedlove, and Peter Raven. 1973. "General Principles of Classification and Nomenclature in Folk Biology." *American Anthropologist* 75:214–42.

Bernal, J. D. 1969. *Science in History*. Vol. 2. London: C. A. Watts.

Bernard, H. Russell, and Pertti Pelto. 1987. *Technology and Social Change.* 2d ed. Prospect Heights, Ill.: Waveland Press.

Bernheimer, Charles, and Claire Kahane, eds. 1985. *In Dora's Case: Freud—Hysteria—Feminism.* New York: Columbia University Press.

Béteille, André. 1981. "The Indian University: Academic Standards and the Pursuit of Equality." *Minerva* 19(2):282–311.

Bijker, Wiebe, and John Law, eds. 1992. *Shaping Technology/Building Society.* Cambridge: MIT Press.

Biology and Gender Study Group. 1989. "The Importance of Feminist Critique for Contemporary Cell Biology." In Nancy Tuana, ed., *Feminism and Science.* Bloomington: Indiana University Press.

Bloor, David. 1976. *Knowledge and Social Imagery.* London: Routledge.

— 1982. "Durkheim and Mauss Revisited: Classification and the Sociology of Knowledge." *Studies in the History and Philosophy of Science* 13(4):267–97.

Bodley, John H. 1982. *Victims of Progress.* 2d ed. Palo Alto: Mayfield.

Boon, James. 1982. *Other Tribes, Other Scribes.* Cambridge: Cambridge University Press.

Borba, Marcelo. 1990. "Ethnomathematics and Education." *For the Learning of Mathematics* 10(1):39–43.

— 1992. "What is New in Mathematics Education? Ethnomathematics: The Voice of Sociocultural Groups in Mathematics Education." *Clearing House.*

Boster, James, Brent Berlin, and John P. O'Neill. 1986. "The Correspondence of Jivaroan to Scientific Ornithology." *American Anthropologist* 88(3):569–83.

Boster, James, and Roy D'Andrade. 1989. "Natural and Human Sources of Cross-Cultural Agreement in Ornithological Classification." *American Anthropologist* 91(1):132–42.

Botelho, Antonio José. 1989. "Struggling to Survive: The Brazilian Society for the Progress of Science (SBPC) and the Authoritarian Regime (1964–1980)." *Historia Scientiarum* 38:45–63.

— 1990. "Far from Silicon Valley: Give Me a Laboratory and I Will Not Raise the World." Paper presented at the annual meeting of the Society for Social Studies of Science, Minneapolis.

Bourdieu, Pierre. 1984. *Distinction.* Cambridge: Harvard University Press.

Boyer, Paul, and Stephen Nissenbaum. 1974. *Salem Possessed: The Social Origins of Witchcraft.* Cambridge: Harvard University Press.

Boxer, Charles R. 1957. *The Dutch in Brazil, 1624–54.* Oxford: Oxford University Press.

Braude, Ann. 1989. *Radical Spirits: Spiritualism and Women's Rights in Nineteenth-Century America.* Boston: Beacon.

Braverman, Harry. 1974. *Labor and Monopoly Capital.* New York: Monthly Review Press.

Brockway, Lucile. 1979. *Science and Colonial Expansion: The Role of the British Royal Botanical Gardens.* New York: Academic Press.

Brokensha, David, D. M. Warren, and Oswald Werner, eds. 1980. *Indigenous Knowledge Systems and Development*. Washington, D.C.: University Press of America.

Brumberg, Joan. 1988. *Fasting Girls*. Cambridge: Harvard University Press.

Bullard, Robert. 1990. *Dumping in Dixie: Race, Class, and Environmental Equity*. Boulder: Westview.

Bulmer, Ralph. 1967. "Why Is the Cassowary Not a Bird?" *Man* 2:1–25.

Burger, Julian. 1987. *Report from the Frontier*. Cambridge: Cultural Survival Press.

Busch, Lawrence. 1991. "Manufacturing Plants: Notes on the Culture of Nature and the Nature of Culture." *International Journal of Sociology of Agriculture and Food* 2:105–15.

Butterfield, Herbert. 1957. *The Origins of Modern Science*. Rev. ed. New York: Free Press.

Callebaut, Werner. 1993. *Taking the Naturalistic Turn, or How Real Philosophy of Science is Done*. Chicago: University of Chicago Press.

Callon, Michel, John Law, and Arie Rip, eds. 1986. *Mapping the Dynamics of Science and Technology*. London: Macmillan.

Canguilheim, Georges. 1988. *Ideology and Rationality in the History of the Life Sciences*. Cambridge: MIT Press.

Caporael, Linnda. 1986. "Anthropomorphism and Mechanomorphism: Two Faces of the Human Machine." *Computers in Human Behavior* 2:215–34.

Caro, Robert. 1974. *The Power Broker: Robert Moses and the Fall of New York*. New York: Random House.

Carroll, P. Thomas, and Lewis Feuer. 1976. "Further Evidence that Karl Marx Was Not the Recipient of Charles Darwin's Letter Dated 13 October 1880." *Annals of Science* 33:385–7.

Carroll, Raymonde. 1988. *Cultural Misunderstandings: The French-American Experience*. Chicago: University of Chicago Press.

Carter, Ruth, and Gill Kirkup. 1990. *Women in Engineering: A Good Place to Be?* Basingstoke: Macmillan.

Certeau, Michel de. 1984. *The Practice of Everyday Life*. Berkeley: University of California Press.

Chapin, Mac. 1991. "How the Kuna Keep Scientists in Line." *Cultural Survival Quarterly* 15(3):17.

Chavkin, Wendy, ed. 1984. *Double Exposure: Women's Health Hazards on the Job and at Home*. New York: Monthly Review Press.

Chernela, Janet. 1988. "Potential Impact of a Proposed Amazon Hydropower Project." *Cultural Survival Quarterly* 12(2):20–24.

— 1990. "The Role of Indigenous Organizations in International Policy Development: The Case of an Awa Biosphere Reserve in Colombia and Ecuador." In Darrell Posey and William Overal, eds., *Ethnobiology: Implications and Applications. Proceedings of the First International Congress of Ethnobiology (Belém, 1988)*. Belém: Museu Paraense Emílio Goeldi.

Clarke, Adele. N.d. *Disciplining Reproduction: Modern American Sciences and the Problem of "Sex."* Berkeley and Los Angeles: University of California Press, forthcoming.

Clarke, Adele, and Theresa Montini. 1993. "The Many Faces of RU4986: Tales of Situated Knowledges and Technological Contestations." *Science, Technology, and Human Values* 18(1):42–78.

Clay, Jason. 1989. "Radios in the Rain Forest." *Technology Review* 92(7):52–57.

— 1990. "Editorial: Genes, Genuis, and Genocide." *Cultural Survival Quarterly* 14(4):1.

— 1992. "Why Rainforest Crunch?" *Cultural Survival Quarterly* 16(2):31–34.

Clifford, James, and George Marcus, eds. 1986. *Writing Culture*. Berkeley and Los Angeles: University of California Press.

Cockburn, Cynthia. 1985. "Caught in the Wheels: The High Cost of Being a Female Cog in the Male Machinery of Engineering." In Donald MacKenzie and Judy Wajcman, eds., *The Social Shaping of Technology*. Philadelphia: Open University Press.

Collins, Harry. 1985. *Changing Order: Replication and Induction in Scientific Practice*. Beverly Hills: Sage.

Collins, Harry, and Trevor Pinch. 1982. *Frames of Meaning: The Social Construction of Extraordinary Science*. London: Routledge.

Committee on Women in Science and Engineering of the National Research Council. 1991. *Women in Science and Engineering: Increasing their Numbers in the 1990s*. Washington, D.C.: National Academy Press.

Condon, Richard. 1989. "Higher Education on Alaska's North Slope." *Cultural Survival Quarterly* 13(1):55–58.

Copenhaver, Brian. 1990. "Natural Magic, Hermetism, and Occultism in Early Modern Science." In David C. Lindberg and Robert S. Westman, eds., *Reappraisals of the Scientific Revolution*. Cambridge: Cambridge University Press.

Cowan, Ruth Schwartz. 1983. *More Work for Mother*. New York: Basic.

— 1985a. "How the Refrigerator Got Its Hum." In Donald MacKenzie and Judy Wajcman, eds., *The Social Shaping of Technology*. Philadelphia: Open University Press.

— 1985b. "The Industrial Revolution in the Home." In Donald MacKenzie and Judy Wajcman, eds., *The Social Shaping of Technology*. Philadelphia: Open University Press.

Crapanzano, Vincent, and Vivian Garrison, eds. 1977. *Case Studies in Spirit Possession*. New York: John Wiley.

Croes, Kenneth. 1994. "Fighting Science with Science: The Cultural Reconstruction of Science and Scientists in the James Bay Hydroelectric Development Controversy." M.S. thesis, Science and Technology Studies Department, Rensselaer Polytechnic Institute.

Cross, Stephen, and William Albury. 1987. "Walter B. Cannon, L. J. Henderson, and the Organic Analogy." *Osiris*, 2d ser., 3:165–92.

Culler, Jonathan. 1976. *Ferdinand de Saussure*. New York: Penguin.

D'Ambrosio, Ubiritan. 1985. "Ethnomathematics and its Place in the History of

Pedagogy of Mathematics." *For the Learning of Mathematics,* vol. 5, no. 1.

DaMatta, Roberto. 1991. *Carnivals, Rogues, and Heroes.* Notre Dame: University of Notre Dame Press.

Dankelman, Irene, and Joan Davidson. 1988. *Women and the Environment in the Third World.* London: Earthscan.

Davis-Floyd, Robbie. 1992a. *Birth as an American Rite of Passage.* Berkeley: University of California Press.

— 1992b. "The Technocratic Body and the Organic Body: Cultural Models for Women's Birth Choices." In David Hess and Linda Layne, eds., *Knowledge and Society Volume 9: The Anthropology of Science and Technology.* Greenwich, Conn.: JAI Press.

Deason, Gary. 1986. "Reformation Theology and the Mechanistic Conception of Nature." In D. Lindberg and R. Numbers, eds., *God and Nature.* Berkeley: University of California Press.

Deerinwater, Jessie, and Sande Smith. 1993. "Native Americans Against Nuclear Waste." *Listen Real Loud* 12(1):8–9.

DeLanda, Manuel. 1992. "Nonorganic Life." In Jonathan Crary and Sanford Kwinter, eds., *Incorporations: Zone 6.* New York: Zone Books.

Derrida, Jacques. 1978. "Structure, Sign, and Play in the Discourse of the Human Sciences." In *Writing and Difference,* pp. 278–93. Chicago: University of Chicago Press.

DeWalt, Billie, and David Barkin. 1987. "Seeds of Change: The Effects of Hybrid Sorghum and Agricultural Modernization in Mexico." In H. Russell Bernard and Pertti Pelto, eds., *Technology and Social Change. Second Edition.* Prospect Heights, Ill.: Waveland Press.

Dhanagare, D. N. 1984. "Equality in the Indian University." *Minerva* 32(3–4): 388–403.

Dibblin, Jane. 1988. *Day of Two Suns.* New York: New Amsterdam Books.

Dickson, David. 1984. *The New Politics of Science.* New York: Pantheon.

Di Nicola, Vincenzo. 1990a. "Anorexia Multiforme: Self-Starvation in Historical and Cultural Context. Part 1: Self-Starvation as a Historical Chameleon." *Transcultural Psychiatric Research Review* 27:165–96.

— 1990b. "Anorexia Multiforme: Self-Starvation in Historical and Cultural Context. Part 2: Anorexia Nervosa as a Culture-Reactive Syndrome." *Transcultural Psychiatric Research Review* 27:245–86.

Doorly, Moyra. 1985. "A Woman's Place: Dolores Hayden on the 'Grand Domestic Revolution.'" In Donald MacKenzie and Judy Wajcman, eds., *The Social Shaping of Technology.* Philadelphia: Open University Press.

Doughty, Paul. 1987. "Against the Odds: Collaboration and Development at Vicos." In Donald D. Stull and Jean J. Schensul, eds., *Collaborative Research and Social Change,* pp. 129–57. Boulder: Westview.

Downey, Gary. 1988. "Structure and Practice in the Cultural Identities of Scientists: Negotiating Nuclear Wastes in New Mexico." *Anthropological Quarterly* 61(1):26–38.

Downey, Gary, Juan Lucena, and Shannon Hegg. N.d. "Weeded-Out: Critical Reflection in Engineering Education." In Gary Downey, Joe Dumit, and Sharon Traweek, eds., *Cyborgs and Citadels: Anthropological Interventions into Techno-humanisms*. Santa Fe: School of American Research Press, forthcoming.

Downey, Gary, Joe Dumit, and Sharon Traweek, eds. N.d. *Cyborgs and Citadels: Anthropological Interventions into Technohumanisms*. Santa Fe: School of American Research Press, forthcoming.

Dreyfus, Hubert, and Paul Rabinow. 1982. *Michel Foucault: Beyond Structuralism and Hermeneutics*. Chicago: University of Chicago Press.

Dubinskas, Frank. 1988. "Janus Organizations: Scientists and Managers in Genetic Engineering Firms." In Frank Dubinskas, ed., *Making Time: Ethnographic Studies of High-Technology Organizations*. Philadelphia: Temple University Press.

Duhem, Pierre. 1982. *The Aim and Structure of Physical Theory*. Princeton: Princeton University Press.

Dumont, Louis. 1977. *From Mandeville to Marx*. Chicago: University of Chicago Press.

— 1980. *Homo Hierarchicus*. Chicago: University of Chicago Press.

— 1986. *Essays on Individualism*. Chicago: University of Chicago Press.

Durkheim, Emile. 1964. *The Division of Labor in Society*. New York: Macmillan/Free Press.

— [1915] 1965. *The Elementary Forms of the Religious Life*. New York: Free Press.

Durkheim, Emile, and Marcel Mauss. 1963. *Primitive Classification*. Chicago: University of Chicago Press.

Edwards, Jeanette, and Sarah Franklin et al. 1993. *Technologies of Procreation: Kinship in the Age of Assisted Conception.* Manchester: Manchester University Press.

Ehrenreich, Jeffrey David. 1990. "Lifting the Burden of Secrecy: the Emergence of the Awá Biosphere Reserve." *The Latin American Anthropology Review* 1(2): 49–54.

Eisenberg, David, Ronald Kessler, Cindy Foster, Frances Norlock, David Calkins, and Thomas Delbanco. 1993. "Unconventional Medicine in the United States." *New England Journal of Medicine* 328(4):246–52.

Elkin, A. P. 1977. *Aboriginal Men of High Degree*. New York: St. Martin's.

Ellenberger, Henri. 1970. *The Discovery of the Unconscious.* New York: Basic.

Elling, Ray. 1981. "Political Economy, Cultural Hegemony, and Mixes of Traditional and Modern Medicine." *Social Science and Medicine* 15A:89–100.

Elliott, Lawrence. 1966. *George Washington Carver*. Englewood Cliffs, N.J.: Prentice-Hall.

Ellison, David. 1978. *The Bio-Medical Fix: Human Dimensions of Bio-Medical Technologies*. Westport, Conn.: Greenwood.

Escobar, Arturo. 1994. "Welcome to Cyberia." *Current Anthropology* 35(3):1–30.

Etkin, Nina L. 1988. "Ethnopharmacology: Biobehavioral Approaches in the Anthropological Study of Indigenous Medicines." In Bernard J. Siegel, Alan R. Beals, and Stephen A. Tyler, eds., *Annual Review of Anthropology*, vol. 17. Palo Alto: Annual Reviews.

Ferguson, Marilyn. 1980. *The Aquarian Conspiracy*. Los Angeles: Tarcher.

Fisher, William. N.d. "Kayapó Environmental Protest and Macro-Development in Pará, Brazil." *Human Organization* 53(4), forthcoming.

Fleck, Ludwik. 1979. *Genesis and Development of a Scientific Fact*. Chicago: University of Chicago Press.

Ford, Richard. 1978. "Ethnobotany: Historical Diversity and Synthesis." In Richard Ford, ed., *The Nature and Status of Ethnobotany*. Anthropological Papers no. 67 Ann Arbor, Mich.: Museum of Anthropology, University of Michigan.

Forman, Paul. 1971. "Weimar Culture, Causality, and Quantum Theory, 1918–1927: Adaptation by German Physicists and Mathematicians to a Hostile Intellectual Envrionment." In Russell McCormach, ed., *Historical Studies in the Physical Sciences*. Philadelphia: University of Pennsylvania Press.

Forsythe, Diana. 1990. "Engineering Knowledge: The Construction of Knowledge in Artificial Intelligence." Technical Report no. CS-90–9, Computer Science Department, University of Pittsburgh.

— 1992. "Blaming the User in Medical Informatics: The Cultural Nature of Scientific Practice." In David Hess and Linda Layne, eds., *Knowledge and Society*. Volume 9. *The Anthropology of Science and Technology*. Greenwich, Conn.: JAI Press.

— 1993. "The Construction of 'Work' in Artificial Intelligence." *Science, Technology, and Human Values* 18(4):460–79.

Fortes, Jacqueline, and Larissa Lomnitz. 1994. *Becoming a Scientist in Mexico*. University Park: Pennsylvania State University Press.

Foucault, Michel. 1970. *The Order of Things*. New York: Random/Vintage.

— 1986. *The Care of the Self: The History of Sexuality*. Volume 3. New York: Random House.

— 1991. "Politics and the Study of Discourse." In Graham Burchell, Colin Gordon, and Peter Miller, eds., *The Foucault Effect: Studies in Governmentality*. Chicago: University of Chicago Press.

Frankenburg, Ronald. 1980. "Medical Anthropology and Development: A Theoretical Perspective." *Social Science and Medicine* 14B:197–207.

Fraser, Nancy. 1991. "Foucault on Modern Power: Empirical Insights and Normative Confusions." In Peter Burke, ed., *Critical Essays on Michel Foucault*. Vol. 2. *Critical Thought Series*. Aldershot, Hants: Scolar Press.

Freire, Paulo. 1970. *Pedagogy of the Oppressed*. New York: Seabury.

Freud, Sigmund. 1962. *The Ego and the Id*. New York: Norton.

Friday, Robert. 1989. "Discussion Patterns in Behaviors of German and American

Managers." *International Journal of Intercultural Relations* 13:429–45.

Friedlander, Edward. 1989. "Dream Your Cancer Away: The Simontons." In Douglas Stalker and Clark Glymour, eds., *Examining Holistic Medicine*. Buffalo: Prometheus Books.

Fry, Tony, and Anne-Marie Willis. 1989. "Aboriginal Art: Symptom or Success?" *Art in America* 77(7):108–17, 159–63.

Fucini, Joseph, and Suzy Fucini. 1990. *Working for the Japanese*. New York: Free Press.

Fuller, Steve. 1993. *Philosophy, Rhetoric, and the End of Knowledge*. Madison: University of Wisconsin Press.

Gaines, Atwood. 1992. *Ethnopsychiatry: The Cultural Construction of Professional and Folk Psychiatries*. Albany: SUNY Press.

Galtung, Johan. 1980. "A Structural Theory of Imperialism." In Johan Galtung, ed., *Peace and World Structure: Essays in Peace Research*. Vol. 4. Copenhagen: Christian Ejlers.

Geisson, Gerald. 1993. "Research Schools and New Directions in the Historiography of Science." *Osiris* 8:227–39.

Gelobter, Michael. 1993. "The Movement for Environmental Justice in the United States." Lecture given for "Non-Western Perspectives on the Environment: The Third Comparative Science and Technology Conference," Rensselaer Polytechnic Institute, Troy, N.Y.

Gelsanliter, David. 1990. *Jump Start*. New York: Farrar, Straus, Giroux.

Genova, Judith. 1989. "Women and the Mismeasure of Thought." In Nancy Tuana, ed., *Feminism and Science*. Bloomington: Indiana University Press.

Gieryn, Thomas. 1983. "Boundary-Work and the Demarcation of Science from Non-Science." *American Sociological Review* 48:781–95.

Gilbert, Scott. 1988. "Cellular Politics: Ernest Everett Just, Richard B. Goldschmidt, and the Attempt to Reconcile Embryology and Genetics." In Ronald Rainger, Keith Benson, and Jane Maienschein, eds., *The American Development of Biology*. Philadelphia: University of Pennsylvania.

Gilbert, Scott, ed. 1991. *A Conceptual History of Modern Embryology*. Volume 7. *Developmental Biology: A Comprehensive Synthesis*. New York: Plenum.

Ginsburg, Faye. 1992. "Indigenous Media: Faustian Contract or Global Village?" In George Marcus, ed., *Rereading Cultural Anthropology*. Durham: Duke University Press.

— 1994. "Culture/Media: A (Mild) Polemic." *Anthropology Today* 10(2):5–15.

Ginsburg, Faye D., and Rayna Rapp, eds. 1995. *Conceiving the New World Order: The Global Politics of Reproduction*. Berkeley and Los Angeles: University of California Press.

Gleick, James. 1987. *Chaos: Making a New Science*. New York: Penguin.

Goodenough, Ward. 1953. *Native Astronomy in the Central Carolines*. Philadelphia: University Museum, University of Pennsylvania.

Goodenough, Ward, and Stephen Thomas. 1987. "Traditional Navigation in the Western Pacific." *Expedition* 29(3):3–14.

Goonatilake, Susantha. 1992a "Biotechnology and the Merged Evolution of Genes and Culture." *Journal of Social and Evolutionary Systems* 15(3):241–48.

— 1992b. "The Voyages of Discovery and the Loss and Rediscovery of 'Others' Knowledge." *Impact of Science on Society* 167:241–64.

Gorden, Raymond. 1987. "The Guest's General Role in the Family." In Louise Luce and Elise Smith, eds. *Toward Internationalism: Readings in Cross-Cultural Communication*. 2d ed. New York: Harper/Newbury House.

Gordon, Colin. 1987. "The Soul of the Citizen: Max Weber and Michel Foucault on Rationality and Government." In Scott Lash and Sam Whimster, eds., *Max Weber: Rationality and Modernity*. London: Allen and Unwin.

Gordon, Deborah. 1988. "Writing Culture, Writing Feminism: The Poetics and Politics of Experimental Ethnography." *Inscriptions* 3/4:7–26.

Gould, Stephen Jay. 1981. *The Mismeasure of Man*. New York: Norton.

Gourman, Jack. 1989. *The Gourman Report*. 5th ed. Los Angeles: National Education Standards.

Gross, G. R. 1970. "The Organization Set: A Study of Sociology Departments." *American Sociologist* 5:25–29.

Gulick, Edward. 1955. *Europe's Classical Balance of Power*. Westport, Conn.: Greenwood.

Gusterson, Hugh. N.d. *Testing Times: A Nuclear Weapons Laboratory at the End of the Cold War*. Berkeley and Los Angeles: University of California Press.

Hahn, R. A. 1984. "Rethinking 'Illness' and 'Disease.'" In E. Valentine Daniel and Judy Pugh, eds., *South Asian Systems of Healing*. Vol. 18. *Contributions to Asian Studies*. Leiden: E. J. Brill.

Hakken, David, with Barbara Andrews. 1993. *Computing Myths, Class Realities: An Ethnography of Technology and Working People in Sheffield, England*. Boulder: Westview.

Hall, A. Rupert. 1983. *The Revolution in Science, 1500–1750*. London: Longman.

Hall, Edward T., and Mildred Reed Hall. 1987. *Hidden Differences*. Garden City, N.Y.: Anchor.

— 1990. *Understanding Cultural Differences*. Yarmouth, Maine: Intercultural Press.

Hamabata, Matthews Masayuki. 1990. *Crested Kumono*. Ithaca: Cornell University Press.

Haraway, Donna. 1989. *Primate Visions*. London and New York: Routledge.

— 1991. *Simians, Cyborgs, and Women*. London and New York: Routledge.

Harding, Sandra, ed. 1993. *The Racial Economy of Science*. New York: Routledge.

Harris, Marvin. 1968. *The Rise of Anthropological Theory*. New York: Crowell.

Harrison, Phyllis. 1983. *Behaving Brazilian: A Comparison of Brazilian and North American Social Behavior*. Rowley, Mass.: Newberry House.

Harvey, David. 1989. *The Condition of Postmodernity*. Oxford: Blackwell.

Harwood, Alan. 1987. *Rx: Spiritist as Needed*. New York: Wiley.

Harwood, Jonathan. 1987. "National Styles in Science." *Isis* 78:390–414.

— 1993. *Styles of Scientific Thought: The German Genetics Community, 1930–1933*. Chicago: University of Chicago Press.

Hauptman, Laurence. 1986. *The Iroquois Struggle for Survival*. Syracuse: Syracuse University Press.

Heath, Deborah, Gerald Glaser, Sandra Gudmendsen, Sue Curry Jackson, Judith Lewish, and Phyllis Rooney. 1991. "STS and Gender." In Steve Fuller and Sujatha Raman, eds., *Teaching Science and Technology Studies: A Guide for Curricular Planners*. Blacksburg, Va.: Virginia Polytechnic Institute.

Hecht, Susanna, and Alexander Cockburn. 1989. *Fate of the Forest: Developers, Destroyers, and Defenders of the Amazon*. New York: Verso.

Heilbroner, Robert. 1972. *The Worldly Philosophers*. New York: Touchstone.

Heims, Steve. 1991. *Constructing a Social Science for Postwar America: The Cybernetics Group, 1946–53*. Cambridge: MIT Press.

Hendry, John. 1980. "Weimar Culture and Quantum Causality." *History of Science* 18:155–80.

Hess, David J. 1991. *Spirits and Scientists: Spiritism, Ideology, and Brazilian Culture*. University Park: Pennsylvania State University Press.

— 1992a. "Disciplining Heterodoxy, Circumventing Discipline: Parapsychology, Anthropologically." In David Hess and Linda Layne, eds., *Knowledge and Society*. Vol. 9. *The Anthropology of Science and Technology*. Greenwich, Conn.: JAI Press.

— 1992b. "Introduction: The New Ethnography and the Anthropology of Science and Technology." In David Hess and Linda Layne, eds., *Knowledge and Society*. Vol. 9. *The Anthropology of Science and Technology*. Greenwich, Conn.: JAI Press.

— 1993. *Science in the New Age*. Madison: University of Wisconsin Press.

— 1994a. "Parallel Universes: Anthropology in the World of Technoscience." *Anthropology Today* 10(2):16–18.

— 1994b. *Samba in the Night: Spiritism in Brazil*. New York: Columbia University Press.

— N.d. "If You're Thinking of Living in STS . . . A Guide for the Perplexed." In Gary Downey, Joe Dumit, and Sharon Traweek, eds., *Cyborgs and Citadels: Anthropological Interventions into Technohumanisms*. Santa Fe: School of American Research Press, forthcoming.

Hess, David J., and Roberto DaMatta, eds. 1994. *The Brazilian Puzzle: Culture on the Borderlands of the Western World*. New York: Columbia University Press.

Hessen, Boris. 1971. *The Social and Economic Roots of Newton's Principia*. New York: Howard Fertig.

Hill, Christopher. 1964. "Puritanism, Capitalism, and Science." *Past and Present* 29:88–97.

— 1975. "Sir Isaac Newton and His Society." In *Change and Continuity in Seventeenth-Century England*. Cambridge: Harvard University Press.

— 1985. *Intellectual Origins of the English Revolution*. London: Oxford University Press.

Hill, Stephen. 1988. *The Tragedy of Technology: Human Liberation Versus Domination in the Late Twentieth Century*. Winchester, Mass.: Pluto Press.

Hodges, Andrew. 1983. *Alan Turing: The Enigma*. New York: Simon and Schuster.

Holmberg, Allan R., Mario C. Vázquez, Paul L. Doughty, J. Oscar Alers, Henry F. Dobyns, and Harold D. Lasswell. 1965. "The Vicos Case: Peasant Society in Transition." *The American Behavioral Scientist* 8(7):3–33.

Holt, Rackham. 1943. *George Washington Carver*. Garden City, N.Y.: Doubleday.

Holton, Gerald. 1988. *Thematic Origins of Scientific Thought*. Rev. ed. Cambridge: Harvard University Press.

Hubbard, Ruth. 1990. *The Politics of Women's Biology*. New Brunswick: Rutgers University Press.

Huff, Toby. 1993. *The Rise of Early Modern Science: Islam, China, and the West*. Cambridge: Cambridge University Press.

Hunter, Michael. 1990. "Science and Heterodoxy: An Early Modern Problem Reconsidered." In David C. Lindberg and Robert S. Westman, eds., *Reappraisals of the Scientific Revolution*. Cambridge: Cambridge University Press.

Hyman, Stanley Edgar. 1974. *The Great Chain of Being: Darwin, Marx, Frazer, and Freud as Imaginative Writers*. New York: Atheneum.

Jackson, Stanley. 1985. "Acedia the Sin and Its Relationship to Sorrow and Melancholia." In Arthur Kleinman and Byron Good, eds., *Culture and Depression*. Berkeley: University of California Press.

Jacob, James. 1977. *Robert Boyle and the English Revolution*. New York: Burt Franklin.

Jacob, Margaret. 1976. *The Newtonians and the English Revolution: 1689–1720*. Ithaca: Cornell University Press.

— 1988 *The Cultural Meaning of the Scientific Revolution*. New York: Knopf.

Jahnige, Paul J. 1989. "Project Letimaren: Indigenous Resource Management in Ecuador's Upper Amazon." *Cultural Survival Quarlery* 13(4):73–74.

Jameson, Frederic. 1984. "Postmodernism, or The Culture Logic of Late Capitalism." *New Left Review* 146:53–92. Also in *Postmodernism; or, The Cultural Logic of Late Capitalism*. 1991. Durham: Duke University Press.

Johnson, Christopher. 1993. *System and Writing in the Philosophy of Jacques Derrida*. Cambridge: Cambridge University Press.

Johnston, William, and Arnold Packer. 1987. *Workforce 2000*. Indianapolis: Hudson Institute, and Washington, D.C.: U. S. Department of Labor.

Jordanova, Ludmilla. 1989. *Sexual Visions*. Madison: University of Wisconsin Press.

Joseph, George. 1991. *The Crest of the Peacock*. London: Tauris.

Just, Ernest Everett. 1936. "A Single Theory for the Physiology of Development and Genetics." *American Naturalist* 70:267–312.

Karp, Herbert, and Sal Restivo. 1974. "Ecological Factors in the Emergence of Modern Science." In Sal Restivo and Christopher Vanderpool, eds., *Comparative Studies in Science and Society*. Columbus, Ohio: Merrill.

Kass-Simon, G., and Patricia Farnes, eds. 1990. *Women in Science: Righting the Record*. Bloomington: University of Indiana Press.

Kauffman, Stuart. 1993. *The Origins of Order: Self-Organization and Selection in Evolution*. Oxford: Oxford University Press.

Keller, Evelyn Fox. 1983. *A Feeling for the Organism*. New York: W. H. Freeman.

— 1985. *Reflections on Gender and Science*. New Haven: Yale University Press.

— 1989a. "The Gender/Science System: or, Is Sex to Gender as Nature is to Science?" In Nancy Tuana, ed., *Feminism and Science*. Bloomington: Indiana University Press.

— 1989b. "Just What *Is* So Difficult about the Concept of Gender as a Social Category?" *Social Studies of Science* 19(4):721–24.

— 1992. *Secrets of Life, Secrets of Death*. New York: Routledge.

Kellert, Stephen. 1993. *In the Wake of Chaos: Unpredictable Order in Dynamical Systems*. Chicago: University of Chicago Press.

Kenny, Michael. 1986. *The Passion of Ansel Bourne*. Washington, D.C.: Smithsonian Institution Press.

Keyes, Charles F. 1985. "The Interpretive Basis of Depression." In A. Kleinman and B. Good, eds., *Culture and Depression*, pp. 152–74. Berkeley: University of California Press

Keynes, John Maynard. 1936. *The General Theory of Employment, Interest, and Money*. New York: Harcourt, Brace.

Kirby, Vicki. 1991. "Comment on Mascia-Lees, Sharpe, and Cohen's 'The Postmodernist Turn in Anthropology: Cautions from a Feminist Perspective.'" *Signs* 16(2):394–400.

Kiste, Robert. 1987. "Relocation and Technological Change in Micronesia." In H. Russell Bernard and Pertti J. Pelto, eds., *Technology and Social Change*. New York: Macmillan.

Kleinman, Arthur. 1980. *Patients and Healers in the Context of Culture*. Berkeley: University of California Press.

— 1985. "Somatization: The Interconnections in Chinese society Among Culture, Depressive Experiences, and the Meanings of Pain." In Arthur Kleinman and Byron Good, eds., *Culture and Depression*. Berkeley: University of California Press.

— 1988. *Rethinking Psychiatry*. New York: Free Press.

Kleinman, Arthur, and Byron Good. 1985. *Culture and Depression*. Berkeley: University of California Press.

Knorr-Cetina, Karin. 1981. *The Manufacture of Knowledge*. New York: Pergamon.

Knorr-Cetina, Karin, and Michael Mulkay, eds. 1983. *Science Observed*. Beverly Hills: Sage.

Kolenda, Pauline. 1976. "Seven Types of Hierarchy in *Homo Hierarchicus*." *Journal of Asian Studies* 35(4):581–96.

Kruse, John, Judith Kleinfeld, and Robert Travis. 1984. "Energy Development on Alaska's North Slope: Effects on the Inupiat Population." In Herbert Applebaum, ed., *Work in Non-Market and Transitional Societies,* pp. 429–90. Albany: SUNY Press.

Kuhn, Thomas. 1970. *The Structure of Scientific Revolutions*. Chicago: University of Chicago Press.

Latour, Bruno. 1987. *Science in Action*. Cambridge: Harvard University Press.

— 1990. "Postmodern? No, Simply *Amodern*! Steps Toward an Anthropology of Science." *Studies in the History and Philosophy of Science* 21(1):145–71.

Latour, Bruno, and Steve Woolgar. 1979. *Laboratory Life: The Social Construction of Scientific Facts*. Beverly Hills: Sage.

Laudan, Larry. 1990. *Science and Relativism*. Chicago: University of Chicago Press.

Lave, Jean. 1988. *Cognition in Practice: Mind, Mathematics, and Culture in Everyday Life*. Cambridge: Cambridge University Press.

Law, John. 1991. "Power, Discretion, and Strategy." In John Law, ed., *A Sociology of Monsters: Essays on Power, Technology, and Domination*. Sociology Review Monograph no. 38. London and New York: Routledge.

Lawson, Michael. 1982. *Damned Indians*. Norman: University of Oklahoma Press.

Layne, Linda. 1992. "Of Fetuses and Angels: Fragmentation and Integration in Narratives of Pregnancy Loss." In David Hess and Linda Layne, eds., *Knowledge and Society* Vol. 9. *The Anthropology of Science and Technology*. Greenwich, Conn.: JAI Press.

Leahey, Thomas. 1992. *A History of Psychology: Main Currents in Psychological Thought*. Englewood Cliffs, N.J.: Prentice Hall.

Lee, Richard E. 1984. *The Dobe !Kung*. New York: Holt, Rhinehart, and Winston.

Lerner, Lawrence, and Edward Gosselin. 1986. "Galileo and the Specter of Bruno." *Scientific American* 255(5):126–33.

Leslie, Charles. 1980. Medical Pluralism in World Perspective. *Social Science and Medicine* 14B(4):191–95.

Lévi-Strauss, Claude. 1961 *Tristes Tropiques*. New York: Atheneum.

— 1963a. *Structural Anthropology*. New York: Basic Books.

— 1963b. *Totemism*. Boston: Beacon.

— 1966. *The Savage Mind*. Chicago: University of Chicago Press.

— 1969. *The Elementary Structures of Kinship*. Boston: Beacon

— 1975. *The Raw and the Cooked*. New York: Harper.

Lewin, Roger. 1992. *Complexity: Life at the Edge of Chaos*. New York: Macmillan.

Lewis, Gilbert Newton, and Merle Randall. 1961. *Thermodynamics*. New York: McGraw-Hill.

Lewontin, R. C., Steven Rose, and Leon J. Kamin. 1984. *Not in Our Genes: Biology, Ideology, and Human Nature*. New York: Pantheon.

Lima, Roberto Kant de. 1992. "The Anthropology of the Academy." In David Hess and Linda Layne, eds., *Knowledge and Society Vol. 9: The Anthropology of Science and Technology*. Greenwich, Conn.: JAI Press.

Lindberg, David. 1990. "Conceptions of the Scientific Revolution from Bacon to Butterfield: A Preliminary Sketch." In David Lindberg and Robert Westman, eds., *Reappraisals of the Scientific Revolution*. Cambridge: Cambridge University Press.

Lindberg, David, and Ronald Numbers. 1986. "Introduction." In David Lindberg and Ronald Numbers, eds., *God and Nature*. Berkeley: University of California Press.

Lindberg, David, and Robert Westman. 1990. *Reappraisals of the Scientific Revolution*. Cambridge: Cambridge University Press.

Lipton, Michael, with Richard Longhurst. 1989. *New Seeds for Poor People*. Baltimore: Johns Hopkins University Press.

Lloyd, G. E. R. 1970. *Early Greek Science: Thales to Aristotle*. New York: Norton.

— 1973. *Greek Science After Aristotle*. New York: Norton.

Lomnitz, Larissa. 1979. "Hierarchy and Peripherality: The Organization of a Mexican Research Institute." *Minerva* 17(4):527–48.

Lovejoy, Arthur O. 1936. *The Great Chain of Being*. Cambridge: Harvard University Press.

Lyotard, Jean-François. 1984. *The Postmodern Condition: A Report on Knowledge*. Minneapolis: University of Minnesota Press.

MacCormack, Carol, and Marilyn Strathern, eds. 1980. *Nature, Culture, and Gender*. Cambridge: Cambridge University Press.

McFarland, Alan. 1970. *Witchcraft in Tudor and Stuart England*. New York: Harper and Row.

McGrayne, Sharon. 1993. *Nobel Prize Women in Science*. New York: Birch Lane Press.

McHale, Brian. 1987. *Postmodernist Fiction*. New York: Methuen.

McIlwee, Judith, and Gregg Robinson. 1992. *Women in Engineering: Gender, Power, and Workplace Culture*. Albany: SUNY Press.

MacKenzie, Donald. 1978. "Statistical Theory and Social Interests: A Case Study." *Social Studies of Science* 8:35–83.

— 1981. "Interests, Postivism, and History." *Social Studies of Science* 11:498–501.

— 1983. *Statistics in Britain*. Edinburgh: University of Edinburgh Press.

— 1984. "Reply to Yearly." *Studies in the History and Philosophy* 15(3):251–59.

McMurray, Linda. 1981. *George Washington Carver*. Oxford: Oxford University Press.

Manber, David. 1967. *Wizard of Tuskegee: The Life of George Washington Carver*. New York: Crowell-Collier.

Mandel, Ernst. 1975. *Late Capitalism*. London: NLB.

Manning, Kenneth. 1983. *Black Apollo of Science*. Oxford: Oxford University Press.

Marcus, George, and Michael Fischer. 1986. *Anthropology as Cultural Critique*. Chicago: University of Chicago Press.

Marglin, Frédérique Apffel. 1990. "Smallpox in Two Systems of Knowledge." In Frédérique Apffel Marglin and Stephen Marglin, eds., *Dominating Knowledge*. Oxford: Clarendon.

Marshall, Alfred. 1961 *Principles of Economics*. New York: Macmillan.

Martin, Brian, C. M. A. Baker, C. Manwell, and C. Pugh. 1986. *Intellectual Suppression*. North Ryde, N.S.W., Australia: Angus and Robertson.

Martin, Emily. 1987. *The Woman in the Body*. Boston: Beacon.

— 1991. "The Egg and the Sperm: How Science Has Constructed a Romance Based on Stereotypical Male-Female Roles." *Signs* 16(3):485–501.

— 1992. "The End of the Body?" *American Ethnologist* 19(1):121–40.

— 1994. *Flexible Bodies: Tracking Immunity in America from the Days of Polio to the Age of AIDS*. Boston: Beacon Press, forthcoming.

— N.d. "Cultural Hybrids." *Cultural Anthropology*. Forthcoming.

Martin, Laura. 1986. "'Eskimo Words for Snow': A Case Study in the Genesis and Decay of an Anthropological Example." *American Anthropologist* 88(2):418–23.

Mascia-Lees, Frances, Patricia Sharpe, and Colleeen Cohen. 1989. "The Postmodernist Turn in Anthropology: Cautions from a Feminist Perspective." *Signs* 15(11):7–33.

— 1991. "Reply to Kirby." *Signs* 16(2):401–8.

Mauss, Marcel. 1967. *The Gift: Forms and Functions of Exchange in Archaic Societies*. New York: Norton.

Merchant, Carolyn. 1980. *The Death of Nature*. San Francisco: Harper and Row.

Merton, Robert. 1970. *Science, Technology, and Society in Seventeenth-Century England*. New York: Howard Fertig.

Meyerhof, Max. 1931. "Science and Medicine." In Thomas Arnold, ed., *The Legacy of Islam*. 1st ed. Oxford: Oxford University Press.

Michaels, Eric. 1986. *The Aboriginal Invention of Television: Central Australia 1982–86*. Canberra: Institute for Aboriginal Studies.

Millroy, Wendy. 1990. "An Ethnographic Study of the Mathematical Ideas of a Group of Carpenters (Spatial Visualization)." Ph.D. diss., School of Education, Cornell University.

Moore, Barrington. 1978. *Injustice: the Social Bases of Obedience and Revolt*. White Plains, N.Y.: M. E. Sharpe

Morain, Genelle. 1987. "Kinesics and Cross-Cultural Understanding." In Louise Luce and Elise Smith, eds. *Toward Internationalism: Readings in Cross-Cultural Communication*. 2d ed. New York: Harper.

Mulkay, Michael, Jonathan Potter, and Steven Yearley. 1983. "Why an Analysis of

Scientific Discourse is Needed." In Karin Knorr-Cetina and Michael Mulkay, eds., *Science Observed*. Beverly Hills: Sage.

Mullings, Leith. 1984. "Minority Women, Work, and Health." In Wendy Chavkin, ed., *Double Exposure: Women's Health Hazards on the Job and at Home*. New York: Monthly Review Press.

Murray, Gerald F. [1987] 1989. "The Domestication of Wood in Haiti: A Case Study in Applied Evolution." In Aaron Podolefsky and Peter J. Brown, *Applying Anthropology*, pp. 148–56. Mountain View, Cal.: Mayfield.

Myers, Fred. 1992. "Representing Culture: The Production of Discourse(s) for Aboriginal Acrylic Paintings." In George Marcus, ed., *Rereading Cultural Anthropology*. Durham: Duke University Press.

Myers, Greg. 1989. "Nineteenth-Century Popularizations of Thermodynamics and the Rhetoric of Social Prophecy." In Patrick Brantlinger, ed., *Energy and Entropy: Science and Culture in Victorian Britain*. Bloomington: Indiana University Press.

Nakane, Chie. 1970. *Japanese Society*. Berkeley: University of California Press.

Nanda, Meera. 1991. "Is Modern Science a Western, Patriarchal Myth? A Critique of the Populist Orthodoxy." *South Asia Bulletin* 11(1/2):32–62.

Nash, June. 1979. *We Eat the Mines and the Mines Eat Us*. New York: Columbia University Press.

Needham, Joseph. 1973. "The Historian of Science as Ecumenical Man." In Shigeru Nakayama and Nathan Sivan, eds., *Chinese Science: Explorations of an Ancient Tradition*. Cambridge: MIT Press.

— 1974. "Science and Society in East and West." In Sal Restivo and Christopher Vanderpool, eds., *Comparative Studies in Society and History*. Columbus: Merrill.

Nelkin, Dorothy. 1982. *The Creation Controversy*. New York: Norton.

Neugebauer, O. 1969. *Exact Sciences in Antiquity*. Dover.

Nietschmann, Bernard, and William LeBon. 1987. "Nuclear Weapon States and Fourth World Nations." *Cultural Survival Quarterly* 11(4):5–7.

Noble, David. 1992. *A World Without Women*. New York: Knopf.

Norberg-Hodge, Helene. 1991. *Ancient Futures: Learning from Ladakh*. San Francisco: Sierra Club.

Nye, Mary Jo. 1993. "National Styles? French and English Chemistry in the Nineteenth and Early Twentieth Centuries." *Osiris* 8:30–49.

Obeyesekere, Gananath. 1985. "Depression, Buddhism, and the Work of Culture in Sri Lanka." In A. Kleinman and B. Good, eds., *Culture and Depression*, pp. 134–52. Berkeley: University of California Press.

Office of Technology Assessment of the U.S. Congress. 1990. *Unconventional Cancer Treatments*. TA-H-405. Washington, D.C.: U.S. Government Printing Office.

Omar, Saleh. 1979. "Ibn al-Haytham's Theory of Knowledge and Its Significance for Later Sciences." *Arab Studies Quarterly* 1:67–82.

Ong, Aihwa. 1987. *Spirits of Resistance and Capitalist Discipline*. Albany: SUNY Press.

Ortner, Sherry. 1974. "Is Female to Male as Nature Is to Culture?" In Michelle Zim-balist Rosaldo and Louise Lamphere, eds., *Woman, Culture, and Society*. Stan-ford: Stanford University Press.

Pacey, Arnold. 1983. *The Culture of Technology*. Cambridge: MIT Press.

— 1990. *Technology in World Civilization*. Cambridge: MIT Press.

Parsons, Talcott. 1968. *The Structure of Social Action*. 2 vols. New York: Free Press.

— 1977. *The Evolution of Societies*. Englewood Cliffs, N.J.: Prentice-Hall.

Parsons, Talcott, and Edward Shils, eds. 1951. *Toward a General Theory of Action*. New York: Harper and Row.

Pavlov, Ivan. 1928. *Lectures on Conditioned Reflexes*. New York: International Pub-lishers.

Paz, Octavio. 1962. *The Labyrinth of Solitude*. New York: Grove Press.

Pelto, Pertti. 1973. *The Snowmobile Revolution*. Menlo Park, Cal.: Cummings.

Pelto, Pertti, and Ludger Müller-Wille. 1987. "Snowmobiles: Technological Revo-lution in the Arctic." In H. Russell Bernard and Pertti J. Pelto, eds., *Technology and Social Change*. New York: Macmillan.

Perrin, Noel. 1979. *Giving up the Gun*. Boston: Godine.

Peterson, John H., Jr. 1987. "The Changing Role of an Applied Anthropologist." In Elizabeth M. Eddy and William L. Partridge, eds., *Applied Anthropology in Amer-ica*, pp. 263–81. New York: Columbia University Press.

Pfaffenberger, Bryan. 1988. "Fetishized Objects and Humanised Nature: Towards an Anthropology of Technology." *Man* 23(2):236–52.

— 1992. "Social Anthropology of Technology." *Annual Review of Anthropology* 21:491–516. Palo Alto: Annual Reviews.

Pickering, Andrew. 1984. *Constructing Quarks*. Edinburgh: Edinburgh University Press.

Pickering, Andrew, ed. 1992. *Science as Practice and Culture*. Chicago: University of Chicago Press.

Pingree, David. 1993. "Review of *The Crest of the Peacock*." *Isis* 84(3):548–49.

Piven, Frances Fox, and Richard Cloward. 1977. *Poor People's Movements*. New York: Vintage.

Pliskin, Karen. 1987. *Silent Boundaries: Cultural Constraints on Sickness and Diag-nosis of Iranians in Israel*. New Haven: Yale University Press.

Porter, Richard E., and Larry A. Samovar. 1991. "Basic Principles of Intercultural Communication." In Larry Samovar and Richard Porter, eds., *Intercultural Com-munication: A Reader*. Belmont, Cal.: Wadsworth.

Porush, David. 1992. "Literature as Dissipative Structure: Prigogine's Theory and the Postmodern 'Chaos' Machine." In Mark Greenberg and Lance Schachterle, eds., *Literature and Technology*. Bethlehem: Lehigh University Press.

Posey, Darrell. 1990. "Intellectual Property Rights: What is the Position of Ethno-biology?" *Journal of Ethnobiology* 10(1):93–98.

— 1991. "Effecting International Change." *Cultural Survival Quarterly* 15(3):29–35.

Posey, Darrell, and William Overal, eds. 1990. *Ethnobiology: Implications and Applications. Proceedings of the First International Congress of Ethnobiology (Belém, 1988)*. Belém: Museu Paraense Emílio Goeldi.

Potter, Elizabeth. 1989. "Modeling the Gender Politics in Science." In Nancy Tuana, ed., *Feminism and Science*. Bloomington: Indiana University Press.

Price, David. 1989. *Before the Bulldozer*. Cabin John, Md.: Seven Locks Press.

Prince, Raymond. 1964. "Indigenous Yoruba Psychiatry." In Ari Kiev, ed., *Magic, Faith, and Healing*. New York: Free Press.

Prince, Raymond, and F. Tcheng-Laroche. 1987. "Culture-Bound Syndromes and International Disease Classification." *Culture, Medicine, and Psychiatry* 11(1): 3–19.

Rabinow, Paul. 1989. *French Modern: Norms and Forms of the Social Environment*. Cambridge: MIT Press.

— 1992a. "Artificiality and Enlightenment: From Sociobiology to Biosociality." In Jonathan Crary and Sanford Kwinter, eds., *Zone 6: Incorporations*. New York: Zone Books.

— 1992b. "Severing the Ties: Fragmentation and Redemption in Late Modernity." In David Hess and Linda Layne, eds., *Knowledge and Society*. Vol. 9. *The Anthropology of Science and Technology*. Greenwich, Conn.: JAI Press.

Rapp, Rayna. 1988. "Chromosones and Communication: The Discourse of Genetic Counseling." *Medical Anthropology Quarterly* 2(2):143–57.

— 1990. "Constructing Amniocentesis: Maternal and Medical Discourses." In Faye Ginsburg and Anna Lowenhaupt Sing, eds., *Uncertain Terms*. Boston: Beacon.

— 1991. "Moral Pioneers: Women, Men, and Fetuses on a Frontier of Reproductive Technology." In Micaela di Leonardo, ed., *Gender at the Crossroads of Knowledge*. Berkeley and Los Angeles: University of California Press.

Rashed, R. 1980. "Science as a Western Phenomenon." *Fundamenta Scientiae* 1:7–21.

Reich, Robert. 1991. *The Work of Nations*. New York: Knopf.

Reingold, Nathan. 1978. "National Styles in the Sciences: The United States Case." In E. G. Forbes, ed., *Human Implictions of Scientific Advance: Proceedings of the XVth International Congress of the History of Science*. Edinburgh: Edinburgh University Press.

Reinharz, Shulamit. 1992. *Feminist Methods in Social Research*. New York: Oxford University Press.

Restivo, Sal. 1979. "Joseph Needham and the Comparative Sociology of Chinese and Modern Science." *Research in Sociology of Knowledge, Sciences, and Art*. Vol. 3. Greenwich, Conn.: JAI Press.

— 1985. *The Social Relations of Physics, Mysticism, and Mathematics*. Boston: D. Reidel.

— 1992. *Mathematics in Society and History: Sociological Inquiries. Episteme*. Vol. 20. Boston and Dordrecht: Kluwer.

Reynolds, David. 1980. *The Quiet Therapies: Japanese Pathways to Personal Growth*. Honolulu: University Press of Hawaii.

Rhodes, Robert. 1986. "Breaking New Ground: Agricultural Anthropology." In Edward C. Green, ed., *Practicing Development Anthropology*, pp. 22–66. Boulder: Westview.

Richards, Evelleen, and John Schuster. 1989a. "The Feminine Method as Myth and Accounting Resource." *Social Studies of Science* 19(4):697–720.

— 1989b "So What's Not a Social Category? Or You Can't Have it Both Ways." *Social Studies of Science* 19(4):725–29.

Richardson, Ken. 1990. *Understanding Intelligence*. Philadelphia: Open University Press.

Ringer, Fritz. 1979. *Education and Society in Modern Europe*. Bloomington: Indiana University Press.

Roberts, V. 1957. "The Solar and Lunar Theory of Ibn al-Shátir: A Precopernican Copernican Model." *Isis* 48:428–32.

Ronan, Colin. 1982. *Science: Its History and Development among the World's Cultures*. New York: Facts on File Publications.

Rosenbaum, Milton. 1980. "The Role of the Term Schizophrenia in the Decline of Diagnoses of Multiple Personality." *Archives of General Psychiatry* 37:1383–85.

Rosser, Sue. 1992. *Biology and Feminism: A Dynamic Interaction*. New York: Twayne.

Rothstein, Frances, and Michael Blim. 1991. *Anthropology and the Global Factory*. New York: Bergin and Garvey.

Rouse, Joe. 1991. "What Are Cultural Studies of Scientific Knowledge?" *Configurations* 1(1):1–22.

Rubin, Gayle. 1993. "Thinking Sex: Notes for a Radical Theory of the Politics of Sexuality." In Henry Abelove, Michèle Aina Barale, and David Halperin, eds., *The Lesbian and Gay Studies Reader*. New York: Routledge.

Sacherer, Janice. 1986. "Applied Anthropology and the Development Bureaucracy: Lessons from Nepal." In Edward C. Green, ed., *Practicing Development Anthropology*, pp. 247–63. Boulder: Westview.

Sahlins, Marshall. 1976. *Culture and Practical Reason*. Chicago: University of Chicago Press.

Sahlins, Marshall, and Elman Service, eds. 1988. *Evolution and Culture*. Ann Arbor: University of Michigan Press.

Saliba, George. 1987. "The Role of Maragha in the Development of Islamic Astronomy: A Scientific Revolution before the Renaissance." *Revue de Synthèse* (July-December), 14(3–4):361 73.

— 1991. "The Astronomical Tradition of Maragha: A Historical Survey and Prospects for Future Research." *Arabic Sciences and Philosophy* 1:67–99.

Sapp, Jan. 1987. *Beyond the Gene: Cytoplasmic Inheritance and the Struggle for Authority in Genetics*. Oxford: Oxford University Press.

Saussure, Ferdinand de. 1959. *General Course in Linguistics*. New York: McGraw-Hill.

Schicor, D. 1970. "Prestige of Sociology Departments and the Placing of New Ph.D.'s." *American Sociologist* 5:157–60.

Schwartzman, Simon. 1991. *A Space for Science: The Development of the Scientific Community in Brazil*. University Park: Pennsylvania State University Press.

Schweber, S. S. 1986. "The Empiricst Temper Regnant: Theoretical Physics in the United States, 1920–1950." *Historical Studies in the Physical and Biological Sciences* 17(1):55–98.

Sclove, Richard. 1993. "Technological Politics as if Democracy Really Mattered: Choices Confronting Progressives." In Michael Shuman and Julia Sweig, eds. *Technology for the Common Good*. Washington, D.C.: Institute for Policy Studies.

Scott, James. 1985. *Weapons of the Weak: Everyday Forms of Peasant Resistance*. New Haven: Yale University Press.

— 1990. *Domination and the Arts of Resistance*. New Haven: Yale University Press.

Segal, Aaron. 1987. *Learning by Doing: Science and Technology in the Developing World*. Boulder: Westview.

Seshagiri, S., and M. Chitra. 1983. *Biodynamic Horticulture: Improvements and Extensions*. Monograph series on "Engineering of Photosynthetic Systems." Vol. 15. Tharamani, Madras: Shri AMM Murugappa Chettiar Research Centre.

Shapin, Steven. 1979. "Homo Phrenologicus: Anthropological Perspectives on an Historical Problem." In Barry Barnes and Steven Shapin, eds., *Natural Order: Historical Studies of Scientific Culture*. Beverly Hills: Sage.

— 1988 "Understanding the Merton Thesis." *Isis* 79:594–605.

Shapin, Steven, and Barry Barnes. 1979. "Darwin and Social Darwinism: Purity and History." In Barry Barnes and Steven Shapin, eds., *Natural Order: Historical Studies of Scientific Culture*. Beverly Hills: Sage.

Shapin, Steven, and Simon Schaffer. 1985. *Leviathan and the Air Pump*. Princeton: Princeton University Press.

Sharp, Gene. 1973. *The Politics of Nonviolent Action*. Vols. 1–3. Boston: Porter Sargent/Extending Horizons Books.

Sharp, Lauriston. 1952. "Steel Axes for Stone-Age Australians." *Human Organization* 11(2):17–22.

Shiva, Vandana. 1989. *Staying Alive: Women, Ecology, and Development in India*. London: Zed Books.

Shiva, V., and J. Bandyopadhay. 1980. "The Large and Fragile Community of Scientists in India." *Minerva* 18(4):575–94.

Shore, Cris. 1992. "Virgin Births and Sterile Debates: Anthropology and the New Reproductive Technologies." [Plus comments and reply.] *Current Anthropology* 33(3):295–314.

Simons, Ronald, and Charles Hughes. 1985. *The Culture-Bound Syndromes*. Boston and Dordrecht: D. Reidel.

Skrabanek, Petr. 1989. "Acupuncture: Past, Present, and Future." In Douglas Stalker and Clark Glymour, eds. *Examining Holistic Medicine*. Buffalo: Prometheus Books.

Slocum, Sally Linton. 1975. "Woman the Gatherer: Male Bias in Anthropology." In Rayna Rapp Reiter, ed., *Toward an Anthropology of Women*. New York: Monthly Review Press.

Snyderman, Mark, and Stanley Rothman. 1988. *The IQ Controversy, the Media, and Public Policy*. New Brunswick: Transaction Books.

Stanley, Autumn. 1983. "From Africa to America: Black Women Inventors." In Jan Zimmerman, ed., *The Technological Woman: Interfacing with Tomorrow*. New York: Praeger.

— 1993. *Mothers and Daughters of Invention*. Metuchen, N.J.: Scarecrow Press.

Star, Susan Leigh. 1991. "The Sociology of the Invisible: The Primacy of Work in the Writings of Anselm Strauss." In David Maines, ed., *Social Organization and Social Processes*. New York: Aldine de Gruyer.

Stemerding, Dirk. 1991. *Plants, Animals, and Formulae*. Twente, The Netherlands: School of Philosophy and Social Sciences, University of Twente.

Stepan, Nancy. 1976. *Beginnings of Brazilian Science*. New York: Science History Publications.

Steward, Julian. 1955. *Theory of Culture Change*. Urbana: University of Illinois Press.

Stone, Allucquère Roseanne. 1992. "Virtual Systems." In Jonathan Crary and Sanford Kwinter, eds., *Zone 6: Incorporations*. New York: Zone Books.

Strathern, Marilyn. 1992. *Reproducing the Future: Anthropology, Kinship, and the New Reproductive Technologies*. New York: Routledge.

Suchman, Lucy. 1987. *Plans and Situated Actions: The Problem of Human-Machine Communication*. Cambridge: Cambridge University Press.

— 1990. "Artificial Intelligence as Craftwork." Paper presented at the annual meeting of the Society for Social Studies of Science.

Talbert, Roy. 1987. *FDR's Utopian Arthur Morgan of the TVA*. Jackson: University Press of Mississippi.

Talbot, Michael. 1991. *The Holographic Universe*. New York: Harper's.

Tannen, Deborah. 1990. *You Just Don't Understand*. New York: Ballantine.

Tannen, Deborah, ed. 1993. *Gender and Conversational Interaction*. Oxford: Oxford University Press.

Taussig, Michael. 1980. *The Devil and Commodity Fetishism in South America*. Chapel Hill: University of North Carolina Press.

— 1987. *Shamanism, Colonialism, and the Wild Man*. Chicago: University of Chicago Press.

Thomas, Keith. 1971. *Religion and the Decline of Magic*. New York: Scribner's.

Tinker, Irene. 1990. *Persistent Inequalities: Women and World Development*. Oxford: Oxford University Press.

Totman, Noel. 1980. "Review of Noel Perrin, *Giving up the Gun*." *Journal of Asian Studies* 39(5):599–601.

Toumey, Christopher. 1991. "Modern Creationism and Scientific Authority." *Social Studies of Science* 21(4):681–99.

Traweek, Sharon. 1988. *Beamtimes and Lifetimes*. Cambridge: Harvard University Press.

— 1992a "Border Crossings: Narrative Strategies in Science Studies and among Physicists in Tsukuba Science City, Japan." In Andrew Pickering, ed., *Science as Practice and Culture*. Chicago: University of Chicago Press.

— 1992b "Storytelling, Jokes, and Erotics: Maintaining Distinctions in High-Energy Physics." Paper presented at the annual meeting of the American Anthropological Association, San Francisco.

— 1993 "An Introduction to Cultural and Social Studies of Sciences and Technologies." *Culture, Medicine, and Psychiatry* 17:3–25.

Treichler, Paula. 1991. "How to Have Theory in an Epidemic: The Evolution of AIDS Treatment Activism." In Constance Penley and Andrew Ross, eds., *Technoculture: Cultural Politics*. Vol. 3. Minneapolis: University of Minnesota Press.

Tuana, Nancy. 1989. *Feminism and Science*. Bloomington: Indiana University Press.

Turkle, Sherry. 1978. *Psychoanalytic Politics*. New York: Basic. 2d ed. (New York: Guilford, 1992).

Turnbull, David, and Helen Watson-Verran. N.d. "Science and Other Indigenous Knowledge Production Systems." In S. Jasanoff, G. Markle, T. Pinch, and J. Petersen, eds. *Handbook on Science, Technology, and Society*. Beverly Hills: Sage, forthcoming.

Turner, Allen C. 1987. "Activating Community Participation in a Southern Paiute Reservation Development Program." In Robert M. Wulff and Shirley J. Fiske, eds., *Anthropological Proxis*. Boulder: Westview.

Turner, Terence. 1991. "Representing, Resisting, Rethinking: Historical Transformations of Kayapó Culture and Anthropological Consciousness." In George Stocking, ed., *Colonial Situations*. Madison: University of Wisconsin Press.

— 1992. "Defiant Images: The Kayapó Appropriation of Video." *Anthropology Today* 8(6):5–16.

Turner, Victor. 1967. *The Forest of Symbols*. Ithaca: Cornell University Press.

Vandermolen, Mimi. 1992. "Shifting the Corporate Culture." *Working Woman* (November), pp. 25–28.

Veith, Ilza. 1965. *Hysteria: The History of a Disease*. Chicago: University of Chicago Press.

Visvanathan, Shiv. 1985. *Organizing for Science*. Oxford: Oxford University Press.

— 1988. "On the Annals of the Laboratory State." In Ashis Nandy, ed., *Science, Hegemony, and Violence*. Delhi: Oxford University Press.

Waldrop, M. Mitchel. 1992. *Complexity: The Emerging Science at the Edge of Order and Chaos*. New York: Simon and Schuster.

Walzer, Michael. 1965. *The Revolution of the Saints*. Cambridge: Harvard University Press.

Warren, Kay, and Susan Bourque. 1991. "Women, Technology, and International Development Ideologies: Analyzing Feminist Voices." In Micaela di Leonardo, ed., *Gender at the Crossroads of Knowledge: Feminist Anthropology in the Postmodern Era*. Berkeley: University of California Press.

Watson-Verran, Helen. 1988. "Language and Mathematics Education for Aboriginal Australian Children." *Language and Education* 2:255–73.

Weber, Max. [1904–5] 1958. *The Protestant Ethic and the Spirit of Capitalism*. New York: Scribner's.

— 1978. *Economy and Society*. Berkeley and Los Angeles: University of California Press.

Webster, Andrew. 1991. *Science, Technology, and Society*. New Brunswick, N.J.: Rutgers.

Webster, Charles. 1986. "Puritanism, Separatism, and Science." In David Lindberg and Ronald Numbers, eds., *God and Nature*. Berkeley: University of California Press.

Weeks, Priscilla. 1992. "Fisher Scientists: The Reconstruction of Scientific Discourse." Manuscript. Revised version of "Managing the Reefs: Local and Scientific Strategies." Paper presented at the 1991 annual meeting of the American Anthropological Association, Chicago.

Westman, R. S., and J. E. McGuire. 1977. *Hermeticism and the Scientific Revolution*. Los Angeles: William Andrews Clark Memorial Library, University of California.

White, Leslie. 1949. *The Science of Culture*. New York: Farrar, Straus, and Cudahy.

— 1987. *Leslie A. White: Ethnological Essays*. Beth Dillingham and Robert Carneiro, eds. Albuquerque: University of New Mexico Press.

Winner, Langdon. 1977. *Autonomous Technologies*. Cambridge: MIT Press.

— 1986. "Do Artifacts Have Politics?" In *The Whale and the Reactor*. Chicago: University of Chicago Press.

— 1993a. "If You Liked Chaos, You'll Love Complexity." *New York Times Book Review* February 14, p. 12.

— 1993b. "Upon Opening the Black Box and Finding It Empty: Social Constructivism and the Philosophy of Technology." *Science, Technology, and Human Values* 18(3):362–78.

Winston, Michael R. 1971. "Through the Back Door: Academic Racism and the Negro Scholar in Historical Perspective." *Daedalus* 100(3):678–719.

Woolgar, Steve. 1981a. "Critique and Criticism: Two Readings of Ethnomethodology." *Social Studies of Science* 11:504–14.

— 1981b. "Interests and Explanation in the Social Study of Science." *Social Studies of Science* 11:365–94.

— 1988a. *Science: The Very Idea*. London: Tavistock.

— 1988b. *Knowledge and Reflexivity*. Ed. Steve Woolgar. Beverley Hills: Sage.

— 1991. "The Turn to Technology in the Social Studies of Science." *Science, Technology, and Human Values* 16:20–50.

Wright, Steve. 1991. "The New Technologies of Political Repression: A New Case for Arms Control?" *Philosophy and Social Action* 17(3–4):31–62.

Yates, Frances. 1972. *The Rosicrucian Enlightenment.* London: Routledge.

— 1964. *Giordono Bruno and the Hermetic Tradition.* Chicago: University of Chicago Press.

Yearley, Stephen. 1982. "The Relationship Between Epistemological and Sociological Cognitive Interests: Some Ambiguities Underlying the Use of Interest Theory in the Study of Scientific Knowledge." *Studies in the History and Philosophy* 13(4):353–88.

Young, Allan. 1982. "The Anthropologies of Illness and Sickness." *Annual Review of Anthropology* 11:257–85. Palo Alto: Annual Reviews.

Zabusky, Stacia. 1992. "Multiple Contexts, Multiple Meanings: Scientists in the European Space Agency." In David Hess and Linda Layne, eds., *Knowledge and Society.* Vol. 9. *The Anthropology of Science and Technology.* Greenwich, Conn.: JAI Press.

— 1994. *Enduring Diversity: The Practice of Cooperation in European Space Science.* Princeton: Princeton University Press.

Index